AF553717

STRESS AMONG PREGNANT WOMEN

By

Ms. C. MOHANA SUNDARI

M.A., M.Phil. (Psychology)
Counsellor
Integrated Counselling and Testing Centre
Southern Railway Hospital
Chennai – 600 011, Tamil Nadu

Editors

Dr. B. PRASAD BABU

M.A., B.Ed., M.Phil., Ph.D. (Psychology)
Rehabilitation Counsellor
Rural Rehabilitation Extension Centre for Handicapped
Thiruvallur, Chennai – 602 001, Tamil Nadu

Dr. DIGUMARTI BHASKARA RAO

M.Sc., M.A., M.A., M.Ed., Ph.D. (Education)
Research Director
R.V.R. College of Education
Guntur – 522 006, Andhra Pradesh
digumartibhaskararao@rediffmail.com

DISCOVERY PUBLISHING HOUSE PVT. LTD.
NEW DELHI-110 002

First Published-2008

ISBN: 978-81-8356-316-1

Published by:

DISCOVERY PUBLISHING HOUSE PVT. LTD.
4831/24, Ansari Road, Prahlad Street,
Darya Ganj, New Delhi-110002 (India)
Phone: 23279245 • Fax: 91-11-23253475
E-mail: dphbooks@rediffmail.com • dphtemp@indiatimes.com
website: www.discoverypublishinghouse.com

Printed at
Arora Offset Press
Laxmi Nagar, Delhi–92

Dedicated

to

Mrs. Panabaka Lakshmi

Health Minister

Government of India

for her

commendable service

to the cause of

public health

PREFACE

Stress is a state of tension created due to hardships in life. It occurs when the demands of an individual exceed his resources. It is more common among women, particularly in pregnant women because of the mores of the society, socio-economic status, education, family support, perception of pregnancy, etc. Understanding the very role of stress in the pregnancy period of women, this study has been undertaken to find out the status of stress and its influencing factors among pregnant women.

Women, majority in number, are facing stress during their pregnancy. The variables namely age, month of pregnancy, working and non-working status, joint or nuclear family and socio-economic status did not show any influence in the levels of stress of pregnant women. But, the number of pregnancy and education did show their influence on stress of pregnant women.

This book will be of use to psychologists and counsellors of pregnant women in attending to their roles suitably. This book will also be useful to special educators and health administrators to perform their functions appropriately.

Dr. D. Bhaskara Rao

Sri Sai Soudha

D-43, S.V.N. Colony

Guntur – 522006

India

CONTENTS

1

INTRODUCTION

"Stress during pregnancy can result in premature and low birth-weight babies. Pregnancy can be a difficult time, especially for low income women."

(North Dakota New Mothers' Survey, 1999)

Health Psychology is a branch of psychology that studies the psychological factors and behaviours which relate to physical and mental health. It studies the effect of behaviour on health and the effect of health issues on behaviour and psychological well-being. It examines behaviour related to wellness, illness, prevention of illness, and diagnosis. It also looks at daily habits such as smoking, diet, exercise and stress coping and its management. Frequently, its interventions focus upon buffering the effect of stress on health by promoting enhanced coping or improved social support utilization.

Science and technology bring improvement in the quality of human life in many ways; it also resulted in many new crises. Crowding, noise pollution, competition, social insecurity, unemployment, violence, loneliness, etc., are all accompaniments of modern living. Stress is an integral part of our lives. While stress is considered a major cause of problems of mental and physical health, its effect is not always undesirable. In fact, stress is a basic ingredient of every human life.

Stress can be described as a pattern or response an organism makes to stimulus event that disturbs the equilibrium and exceeds a person's ability to cope with. The stimulus events include a large variety of external and internal conditions called stressors, if they are perceived to threaten one's well-bring and demand some kind of adaptive response.

Stress is an inescapable part of modern life. That's the bad news. The good news is that stress isn't altogether bad news. In metered doses, it can be helpful. It can even make you better at what to do, and help give you the competitive edge. Stress is an adaptive response. It is the body's reaction to an event that is seen as emotionally disturbing, disquieting, or threatening. When we perceive such an event, we experience what is called the fight or flight response. To prepare for fighting or fleeing, the body increases its heart rate and blood pressure; more blood is then sent to the heart and muscles, and the respiration rate increases. This response was probably beneficial to our cavemen ancestors who had to fight off wild animals. But today, stress itself has become the wild animal. Untamed all allowed to run rampant in our lives, it can destroy our health.

This is an age of stress and every individual fees the distress if he or she is unable to manage it. Though stress is universal, some people face more stressors than others. For example in our society women, adolescents, old people, city dwellers, people who are in certain jobs like professional deriving, police, salesmen, business executives, surgeons and people who live in unfavourable and unhealthy environment.

Stress in psychology is used in at least two different ways. First it is defined as the state of psychological upset or disequilibrium in the human beings caused by frustrations, conflicts and other internal as well as external strains and pressures. In a more serious condition of the stress, the individual reaches a point where the physical processes are seriously affected and mental processes are confused, and the emotional state is chaotic.

Psychologists and all others who are committed to facilitating women's career development must ensure that career women should be trained to evolve strategies to help them in identifying constructive solutions to deal with the problem of stress. There is a strong need for additional research on interventions to reduce work stress and ill health among employed women. Since the stress afflictions are becoming more rampant in the working population, it is time now to focus attention to the area of executive health. We have to take a 'holistic health' perspective which will take care of physiological, psycho emotional and also spiritual dimensions.

Stress can arise for a variety of reasons. Stress can be brought about by a traumatic accident, death or emergency situation. Stress can also be a side effect of a serious illness or disease. There is also stress associated with daily life, the workplace, and family responsibilities. It is difficult to stay calm and relaxed in hectic lives. Women, in this generation, have many roles to play, namely, spouse, mother, caregiver, friend, and/or worker. With all the roles that a woman has to play in daily life, it seems almost impossible to find ways to destress. But it's important to find those ways. Ones health depends on it.

Stress can take on in different forms, and can contribute to symptoms of illness. Common symptoms include headache, sleep disorders, difficulty in concentration, short-temper, upset stomach, job dissatisfaction, low morale, depression, and anxiety. We all deal with stressful things like traffic, arguments with spouses, and job problems. Some researchers think that women handle stress in a unique way by tending and befriending.

Everyone has, at one time or other, stress. We have short-term stress, like getting lost while driving or missing the bus. Even everyday events, such as planning a meal or making time for errands, can be stressful. This kind of stress can make us feel worried or anxious.

During stress, women tend to care for their children and find support from their female friends. Women's bodies make chemicals that are believed to promote these responses. One of these chemicals is oxytocin, which has a calming effect during stress. This is the same chemical that is released during childbirth and found at higher levels in breastfeeding mothers, who are believed to be calmer and more social than women who don't breastfeed. Women also have the hormone estrogen, which boosts the effects of oxytocin. Men, however, have high levels of testosterone during stress, which blocks the calming effects of oxytocin and causes hostility, withdrawal, and anger.

The woman conceives that she has to give birth to a male child, otherwise things will be different. Many being ignorant and illiterate feel a lot of physical and emotional stress during pregnancy. Largely, due to neglect, more than 50,000 women, at least 90 percent of them are from the developing world, die each year in pregnancy or after childbirth – an average of one a minute. The astonishing contrast between the rich and the poor nations is that 600 mothers per 1,00,000 live births die in Africa, 400 in Asia, 300 in Latin America and 10 in Northern Europe and North America. In other words, the change of maternal mortality in Africa is one in 20; in the US it is one in 6,366.

This large gap between women in the developing world and in industrialized society is due to lack of adequate medical care, neglect of female children, early marriages, poverty, lack of female education and illiteracy, overwork and under feeding. After the childbirth, her child is her world. She got to devote all her energies and concentration for the family, husband and children. There is no room for her to be considered.

MEANING OF STRESS

Stress is "a state of tension that is created when a person responds to the demands and pressures that come from work, family and other external sources, as well as those that are internally generated from self imposed demands, obligations and self-criticism".

Hans Selye first introduced the concept of stress in 1939. Derived from Latin, the word "stress" was popularly used in the seventeenth century to mean hardship, strait, adversity or affliction.

According to Selye (1956), "any external event or internal drive which threatens to upset the orgasmic equilibrium is stress". He has defined stress as the non-specific response of the body to any demand made upon it.

Lazarus (1960) maintains that "stress occurs when there are demands on the person which tax or exceed his adjustment resources".

FACTORS OF STRESS

The following factors make the individuals more vulnerable to stress.

A. *Individual Factors*

1. *Type a personality:* They are ambitious, always in a hurry, want to achieve more in a short time. They set high targets and quality for themselves.
2. *Low self esteem:* They underrate themselves and have negative attitude and approach.
3. Loneliness and feelings of isolation and neglect.
4. Inadequate knowledge and skills.
5. Health problems.
6. Needs are not fulfilled (food, shelter, sex, love, recognition, etc.).
7. Unhealthy habits.

B. *Family Factors*

1. Confusing and demanding roles, high expectations.
2. Interpersonal relationships – strained with natural distrust, suspicion, hatredness or fear.
3. Inadequate resources of both men and matters.
4. Improper use of reward and punishment.
5. Lack caring love and affection.

C. Job Factors

1. Less or more work, unrealistic targets.
2. Lack of guidance and support from superiors, colleagues and subordinates.
3. Nature of job not clear and job satisfaction.
4. Frequent transfers.
5. Discrimination in giving increments and promotions.
6. Criticism and neglect.

D. Finance and Other Resources

1. Inadequate money and materials.
2. Poor management/lot of restrictions.
3. Big tasks and responsibilities in money matters.

E. Socio-cultural Factors

1. Isolation or over crowding.
2. Severe competition, insecurity, disparities.
3. Disorganization and uncertainties.
4. Social evils and all kinds of exploitation.
5. Un-ethical practices and poor moral standards.
6. Lawlessness.

TYPES OF STRESS

There are three types of stress that are described here under.

1. Conflicts

Every day we face different types of conflicts. We get torn between good and right things (which are difficult and not giving immediate benefits) and bad and wrong things (which are attractive and give immediate benefits). We may have to choose one among many and we are not sure which the best is. We may have to make one – difficult choice among many unpleasant ones and we are not very sure which is least troublesome to do or not to do is the main conflict. There may be inner conflicts between our basic needs (id), reality (ego) and moral and social norms (super-ego). Conflicts may be between two people like husband and wife or between family members or friends and colleagues or individual and society.

2. *Frustrations*

Failures or frustrations are common in our life. Our aspirations and our achievements do not tally. 'We propose and God disposes'. Repeated small failures or occasional big failure can become very stressful. When we fail but our friends/relatives/colleagues succeed, we feel more distressed. When our needs are not fulfilled, we get frustrated.

3. *Pressures*

There may be internal pressure to achieve more earn more, to go up in the socio-economic ladder, there may be a severe anger for recognition name and fame, for money and materials benefits. Family members, friends, relatives and the system want us to achieve more and more. This can be very stressful.

EFFECT OF STRESS

The following are effects of stress.

1. *Heart and Blood Circulation*

We feel the heart pounding in the chest (palpitation). Blood flow increases under the skin and we blush. We feel the warmth. We sweat. The blood flow to the brain is decreased. We feel giddy and in a severe condition we may faint. Blood pressure rises.

2. *Lungs and Respiratory System*

There is an increased but shallow breathing. We may struggle to breath. There may be sensation of constriction in the chest.

3. *Muscles and Joints*

There are contractions of the muscles. There may be shivering. Because of continuous contractions, there may be aches and pains, headache, chest pain, back pain, pain in the limbs, joint pains, easy fatigability are common complaints of people who are under stress.

4. *Genitals and Urinary System*

There may be increased frequency of passing urine, premature ejaculation, no or poor erection of penis. No desire to have sex. Thus a person under stress, may suffer from poor sexual drive and ability.

5. *Digestive System*

Dryness of mouth, poor or no appetite, lack of taste, fullness of stomach, nausea, vomiting, indigestion, increased flatulence, diarrhoea or constipation. Thus food intake becomes irregular and 'not enjoyable'.

6. *Brain and Mind*

Negative emotions like fear, anger, and sadness engulf the individual, decreased attention, concentration, poor memory, inability to take quick decisions, poor learning, and inability to remain comfortable and composed are the common complaints of people under stress. Thus there is a general decline of mental including functioning of the individual. This naturally increases the stress and it becomes a vicious circle.

All these changes occur because of excess secretion of adrenaline as Hypothalamus – Pituitary – Adrenal glands become hyperactive under stress. Prolonged stress leads to increased 'wear and tear' of the body and mind and in turn lead to:

(i) *Accident Proneness*

Because of poor attention, concentration, faulty perception and inappropriate decisions, the individual become more accident prone. People under stress have more road accidents, home or work place accidents than people who are not under stress.

(ii) Mental Disorders

Anxiety disorders (generalized anxiety, panic, phobic disorders) depression, hysteria, somatoform disorders acute psychosis, exacerbation of already existing mental disorders are frequently seen in people who are under stress.

(iii) Alcohol and Other Substance Abuse

It is a common experience of many people that when they are under tension, they smoke and drink more than usual. They run a high risk of abusing these substances and eventually become dependent on them.

(iv) Physical Disorders

Compared to other people, people under stress are more prone to develop:

(i) Infections: Immunity is reduced; disease producing organisms like bacteria, virus, fungi take upper hand and cause infections. Repeated soar throats, skin – boils, upper respiratory tract infections, pneumonia, typhoid, tuberculosis, urinary tract infections, diarrheas and dysentery are common.

(ii) Acid – peptic disorders and peptic ulcers

(iii) Chronic diarrheas and dysenteries (irritable bowel syndrome, uncreative colitis)

(iv) Heat related diseases like hypertension (High BP), Myocardial infarction (Heat attacks).

(v) Chronic bronchitis and asthma

(vi) Chronic arthritis (Joint swelling and pains)

(vii) Menstrual irregularities

(viii) Migraine and tension headaches

(ix) Skin diseases like eczema, psoriasis.

COPING SKILLS

We have to increase our coping skills to manage the stressful issues and situations.

The coping skills of stress are:

1. Ability to relax and remain calm on composed in times of stress.
2. Ability to understand the nature of the problems and think of possible solutions.
3. Maintain self esteem and take control of the situation.
4. Ability to set realistic objectives and goals and try to achieve them.
5. Ability to have more realistic and appropriate attitude, knowledge and changes the behaviour as required by the environmental/situation.
6. Ability to get the help of family members and others in facing the situation or the problem.
7. Ability to have positive thinking and remain optimistic.

HOW TO DEVELOP COPING SKILLS?

Coping skills can be developed through these techniques.

1. *Self-introspection*

Know your assets and limitations: Know your strength and weaknesses, Know your resources is terms of knowledge, money, materials, feel proud of your assets, stop worrying about your weakness and limitations. Plan and try to improve yourself and reduce your limitations. Do not compare yourself with people who are better than you. Look at people below you. Periodically assess your actions and reactions, make necessary changes to suit *yourself and others.*

2. *Cultivate People*

Stop criticizing others, stop finding fault with others, show respect to elders and love to the youngsters. Cultivate friendship. Request people to support you to help you and help you in time of stress. Do not hesitate to ask for help when you know that you cannot handle the problem or situation alone.

Expect less from your family members, friends and colleagues, relatives, try to give more to them. In turn, they will come to your help and rescue when you are in trouble.

3. *Play Your Role Properly*

Each one of us have to play different roles in our family, occupations and social life – as parent, as spouse, as brother or sister, as employer, as employee, as a friend and so on. Understand the role – responsibilities and make an honest effort fit into the role as expected in your community.

4. *Discipline Yourself and Manage Your Resources Well*

Whether it is time, money, materials, every one of us have constraints. None of us have the luxury of having unlimited resources. Therefore, we have to plan, prioritize our needs and allot time, money and materials accordingly. For example if you are earnings Rs. 10,000/- a month, you must allot the money for basic needs like food, clothier, water and electricity bills, rent, school, college fees of the children. Later, if you save money, then only think of spending for luxury items, gifts, of others, celebrations, outing, etc. Make a time table and save time f or essential activities. Keep your room, house, work place neat and tidy. Keep all your things in an order and at proper places so that you can avoid wasting precious time in searching for them.

5. *Understand the Problem and Situation*

Before you plan, act or react to them. Try to know-how and when the problems started, who has contributed to it. What are the aggravating factors and what could be the outcome?

6. *Positive Attitude*

Tell yourself that you will do your best and stop worrying about the results. Keep trying and keep working. Be optimistic and tell yourself that you will succeed; you will be able to manage and sail through the problems.

7. *Set Attainable Goals, Make Action Plans*

Set goals which are practicable and realistic. Keep pursuing your goals in between, periodically look at you gains and feel happy. This will improve your self-esteem.

8. Stop Worrying When Thing Go Wrong

Do not worry about past or future, live for the present. Learn from pat mistakes or other's mistakes. Make appropriate changes in your attitudes, work plans so that you do not repeat the mistakes.

9. Face One Problem at a Time

If you have many problems in hand, make a list of these problems, thin which one you have to handle first. Handle one problem at a time. Once you manage it, take the seconds. Involve others in this process so that they handle some of the problems and help you.

10. Share Your Failures Frustrations

Share your failures, frustrations, negative emotions with your family members and intimate people. You feel the relief. Suppressed feelings are painful and make you very unhappy and uncomfortable. Ventilation helps you to feel good and comfortable.

11. Shun the Superman and Superwoman Image of Yourself

You need not be an Abhimanyu' or Arjuna. You are an ordinary person. You need not take the entire load on your shoulders.

12. Relaxation

In between you busy – schedule and work, find a few minutes to relax. Stretch yourself in the chair. Take a deep and slow breath ten times. Think of 'happy moments' of your life or bringing images of people you like or love in front of you. Look at flowers, plants, trees, birds or animals or children enjoy the nature's creations. Music, arts, light reading, chitchat with friends, colleagues, a cup of fruit juice, milk cool water, will help you to relax.

13. Physical Activities

Every morning and evening spend an hour or so in activities like sports, exercises, yoga, meditation, gardening and other type of physical activities.

14. *Help Others*

Do well to others through a kind word, a good gesture. Give an occasionally, donate something you can afford to, involve yourself in some service activity. Have a nice feeling of doing some thing good to others. This will help you to find a meaning and a purpose to live which increases yourself confidence.

15. *Keep Yourself Away*

Keep away from tobacco, alcohol sedative, hypnotics and other harmful substances.

16. *Take Good Food*

Eat food rich in protein, vitamins, and minerals. Eat at regular intervals. Make eating time relaxing and enjoyable. This is possible, if you have food along with people you like and in a healthy environment. Stop eating, excess sweets / fatty oily items. Stop eating in between the meals.

17. *Lead A Simple and Continual Life*

Do not go behind luxuries and materialistic life. Simple living is environment-friendly also.

18. *Develop Interest in Cultural and Spiritual Activities*

Have belief in the existence of a super force (You may call it as good or something else). This will help you to pay a good hope at the time of stress. Reduce your selfishness and work for the betterment of others.

19. *Have a Family Doctor*

Go to him periodically. Take are of any ailment as soon as it appears. If you are having a chronic health problem, follow the doctor's instructions carefully. See how best you can control the symptoms and minimize the suffering.

20. *Improve the Following 'Life Skills'*

These life skills will help in reducing stress in pregnant women.

(i) Decision-Making

(a) Learning basic steps for decision-making

(b) Making difficult decision

(c) Decisions making about important life plans

(ii) Problem Solving

(a) Basic steps for problem solving

(b) Generating new ideas about things that are taken for granted

(c) Adapting to changing social circumstances

(iii) Creative Thinking

(a) Developing capacities to think in creative ways

(b) Generating new ideas about things that are taken for granted

(c) Adapting to changing social circumstances

(iv) Critical Thinking

(a) Learning he basic processes in critical thinking

(b) Naming objective judgments about choices and risks

(c) Resisting media influence on attitudes towards smoking and alcohol, buying things unnecessarily, choosing a life style which is unhealthy

(v) Effective Communication

(a) Basic verbal and non-verbal communication, skills, communicates freely with family and friends

(b) Assertive communications in the face of peer pressure or social pressure

(c) Using assertiveness to resist pressure to do health damaging activities

(vi) Interpersonal Relationship Skill

(a) Learning

(b) Forming new relationships and surviving loss of friendships

(c) Seeking support and advice from others in a time of needs

(vii) Self Awareness

(a) Learning about oneself as a special person

(b) Self control

(c) Understanding one' right and responsibilities

(viii) Empathy

(a) Understanding how people alike and how they differ and learning to appreciate the differences between people

(b) Avoiding prejudice and discrimination people

(c) Caring of people

(ix) Coping with Emotions

(a) Identifying sources of stress

(b) Methods for coping in stressful situations

(c) Coping with situations of adversity

(d) Learning to express emotions in a socially acceptable manner. Use dramatics, music, dance, painting, writing to express one's emotions

PRENATAL DEVELOPMENT

Prenatal development is divided into three trimesters. During the first two months the developing human is referred to as an embryo. The embryo has three layers from which all body organs develop. During the second trimester the developing human is referred to as a fetus. During the third trimester, the individual is a baby which if born prematurely could survive with extra support. Premature births prior to the third trimester are less likely to survive, even with extraordinary medical care in a neonatal intensive care unit.

First Trimester

(i) First Month

Fertilization, descent of ovum from tube to uterus. Early cell division and formation of embryonic disc from which new organism will develop. Early formation of three layers of cells: (1) the ectoderm, from which sense organs and nervous system will develop; (2) the mesoderm, from which circulatory, skeletal and muscular systems will develop; (3) the endoderm, from which digestive and some glandular systems will develop. Special layer of cells formed in the uterus which will become the placenta and through which nutritive substances will be carried to the new organism and waste products carried away. Special layer of cells forms the amnion or water-sac, which will surround the developing embryo except at umbilical cord. Heart tube forms and begins to pulsate and force blood to circulate through blood vessels in embryonic disc. Nervous system begins to arise, first in form of neural groove. Development of intestinal tract, lungs, liver and kidneys begins. By end of one month, the embryo is about one-fourth inch long, curled into a crescent, with small nubbins on sides of body indicating incipient arms and legs.

(ii) Second Month

Embryo increases in size to about 1½ inches. Bones and muscle begin to round out contours of body. Face and neck develop and begin to give features a human appearance. Forehead very prominent, reflecting precocious development of brain in comparison to rest of body. Limb buds elongate. Muscles and cartilage develop. Sex organs begin to form.

(iii) Third Month

Beginning of fetal period. Sexual differentiation continues, with male sexual organs showing more rapid development and the female remaining more neutral, buds for all 20 temporary teeth laid down. Vocal cords appear; digestive system shows activity. Stomach cells begin to secrete fluid; liver pours bile into intestine. Kidneys begin functioning, with urine gradually seeping into amniotic fluid. Other waste products passed through placenta into mother's blood. Bones and muscles continue development, and by end of third month spontaneous movements of arms, legs, shoulders and fingers are possible.

Second Trimester

(iv) Fourth Month

Lower parts of body show relatively accelerated rate, so that head size decreases from one-half to one-fourth of body size. Back straightens, hands and feet are well-formed. Skin appears dark red, owing to coursing of blood showing through thin skin and wrinkles, owing to absence of underlying fat. Finger closure is possible. Reflexes become more active as muscular maturation continues. Fetus begins to stir and so thrust out arms and legs in movements readily perceived by the mother.

(v) Fifth Month

Skin structures begin to attain final form. Sweat and sebaceous glands are formed and function. Skin derivatives also appear – hair, nails on fingers and toes. Bony axis becomes quite straight and much spontaneous activity occurs. Fetus is lean and wrinkled, about one foot long and weighs about one pound.

(vi) Sixth Month

Eyelids which have been fused shut since third month, reopen; eyes are completely formed. Taste buds appear on tongue and in mouth and are, in fact, more abundant than in the infant or adult.

Third Trimester

(vii) Seventh Month

Organism capable of independent life from this time on. Cerebral hemispheres cover almost the entire brain. Seven-month fetus can emit a variety of specialized responses. Generally is about 15 inches long and weighs about three pounds.

(viii) Eighth and Ninth Month

During this time, finishing touches are being put on the various organs and functional capacities. Fat is formed rapidly over the entire body, smoothing out the wrinkled skin and rounding out body contours. Dull red colour of skin fades so that a firth pigmentation of skin is usually very slight in all races. Activity is usually great and he can change his position within the somewhat crowded uterus. Periods of activity will alternate with periods of quiescence. Fetal organs step up their activity. Fetal heart rate becomes quite rapid. Digestive organs continue to expel more waste products, leading to the formation of a fetal stool, called the meconium, which is expelled shortly after birth. Violent uterine contractions begin, though milder ones have been tolerated earlier, and the fetus is eventually expelled from the womb into an independent physiological existence.

STRESS AND DEPRESSION DURING PREGNANCY

Depression or mental stress is a psyche condition that inhibits personal happiness and productivity during pregnancy. Moderate ("normal") depression or mental stress is a general part of survival. Ranging from person to person it may be minor to major depending in its impact and intensity. Unless persists, depression or moderate stress is a normal and unavoidable phase in our life like tears in eyes. Normally we get depressed and stressed when something bad happens, such as the death of a loved one or an end to a relationship. Usually depression alleviates over a period of time; but in some cases it persists for much longer time. On accumulation for longer period, depression or stress can erode family relationships, and may promote other physical and mental health problems. It occurs in kids and adults. Nowadays, excessive depression has become common in people due to today's unique surrounding. Depression may be due to hormonal (chemical) imbalance, or as a result of psychological "upset." Anti-depression medications can reduce the symptoms to tolerable, but leaves the root cause/ s of depression unhealed.

Even during pregnancy, many women who have had depression may be continuing antidepressants to prevent symptoms. If she is getting treatment under expert's guidance and taking those antidepressants, for which there is a fair amount of information suggesting that it is safe, it is fine to continue. However, if she is taking a medication for which there is little information, she should switch onto a medication thought to be safer.

However most the depression symptoms are mild and short lived; it may be more accurate to consider them as a normal experience rather than a psychiatric illness. In case of mild depression, the experts would recommend trying to treat her symptoms with mild antidepressants. However the first trimester (12 weeks) of pregnancy is a crucial time when medication can cause malformations of the fetus. So experts may advice discontinuing medications over this period unless a woman has had multiple episodes of severe *stress and depression*. After three months, there is not the concern about medication causing malformation of organs. However, there are still questions about whether medication might cause a miscarriage or subtle changes in the early development of the future child.

Whether the woman remains on medication or not, it is always better to opt for some alternative way of healing that may help to prevent recurrence of Depression and stress during pregnancy.

Following are some of common symptoms that indicate 'Emotional Imbalance' during pregnancy:

- Persistent longings.
- Unreasonable longing desires.
- Delusions or hallucinations.
- Noticeably depressed mood however if this depression persists without any interval between subsequent periods then it is more likely that she has another kind of mood problem.

- Blank depression.
- Irritability.
- Get depressed over trivial issues.
- Sadness.
- Aversion.
- Disgust or gross feeling.
- Fear of miscarriage.
- Resentful behaviour.
- Swinging moods.
- Marked anxiety.
- Feeling tensed.
- Feeling twitchy.
- Marked shifts in mood (suddenly tearful, overly sensitive).
- Persistent, marked anger or irritability, increased conflicts.
- Loss of interest in usual activities (e.g., work, hobbies).
- Difficulty concentrating and focusing attention.
- Marked lack of energy, feeling very easily exhausted.
- Noticeable change in appetite, overeating, food cravings or aversion to food.
- Sleeping too much or having a hard time sleeping.
- Feeling overwhelmed or out of control.

Many types of antidepressants are available in market. Usually experts recommend a kind of antidepressant that increases brain levels of a chemical called serotonin. These medications are called selective serotonin reuptake inhibitors

(SSRIs). The SSRIs preferred by the experts for use in pregnancy are fluoxetine (Prozac), sertraline (Zoloft), and paroxetine Paxil). The experts also favour tricyclic antidepressants, another kind of antidepressant that affects other brain chemicals in addition to serotonin. SSRIs may cause the following side effects: nervousness, drowsiness, insomnia, restlessness, nausea, diarrhoea, and sexual problems.

A mother's emotions during pregnancy, especially stress effect her heart rate and blood pressure. These changes also affect the fetus in the short term, according to a new study. The study appears in this month's issue of Developmental and Behavioural Pediatrics.

Previous studies have shown that stress during pregnancy can increase the risk of low birth weight and premature births, says lead researcher Catherine Monk, Ph.D., with the psychiatry department at Columbia University in New York.

A growing body of evidence also suggests that pregnancy stress may "reprogram" the fetal development in ways that affect the baby's behaviour and functioning later in life, she says. For example, babies of women suffering from depression are more difficult to soothe shortly after birth. They also have poorer reflexes and higher levels of the stress hormones cortisol and epinephrine. Whether these changes affect the child's fetal and long-term health and development are not known, says Monk. In the current study, 32 healthy women all in their third trimester took a test designed to produce a stress response. Before, during, and after completing the test, researchers monitored the women's heart rate, blood pressure, and breathing rate. The women also completed a questionnaire that measured their everyday levels of anxiety.

Researchers found that during the stress test, increases in fetal heart rate were indeed related to the mother's overall level of anxiety. However, the fetal heart rate increase did not occur *during* the mother's increase in heart rate or blood pressure. Rather, it occurred after the mother's test, during the recovery period.

This delayed reaction likely occurs because the physical effects of stress need time to reach the fetus, authors say. However, it could also mean that during the previous months of pregnancy, a woman's moods have already shaped the fetus' heart rate patterns, they say irritability and dissatisfaction.

PRENATAL MATERNAL STRESS

The important psychological hazard associated with the prenatal period is maternal stress-heightened general emotionally over a prolonged period of time. Stress can be the result of fear, anger, grief, jealousy, and envy.

There are many causes of maternal stress during pregnancy. The most common of which are the following: not wanting a child because of marital or economical difficulties or because having a child will interfere with educational or vocational plans; physical discomforts that are severe and frequent enough disturbed, feelings of inadequacy for the prenatal role: and fears that are often heightened by mass media reports of the frequency of birth defects and the specific causes of birth defects, such as rubella and thalidomide. Some women have fantasies and dreams about giving birth to deformed babies, which intensifies such fears.

Maternal stress affects the developing child both before and after the birth. Before birth, severe and persistent glandular imbalance due to stress may result in irregularities in the developing child and complications are greater because the infant must often be delivered by instruments. Furthermore anxiety often leads to overeating and excessive weight gain in pregnancy, which further complicates birth.

Sporadic, less severe maternal stress is less likely to lead to developmental irregularities, though it does increase fetal activity. If this increase is slight, the effect is favourable because the fetus will get the exercise necessary for healthy muscular development. If stress leads to excessive fetal activity, however the fetus will be underweight and nervous to the point where early post-natal adjustments will be seriously affected.

Although relatively few studies have been made to the effects of maternal stress on the child after birth, it has been shown that when prolonged, emotional strain affects endocrine balance, anxieties may occur over into the period of the newborn and seriously affect adjustments to post-natal life. The infant may show hyperactivity, which prevents its adjusting to feeding and sleeping patterns, or it may cry excessively.

It has been noted that newborn infants and young babies who were most active, as fetuses show certain motor performances at an earlier age than those who were less active. On the other hand excessive fetal activity causes infants to be considerably underweight and often to be slower in acquiring skills post-natal, a result of excessive nervousness. In addition, as sontag has reported, they often become "hyperactive, irritable, squirming, crying" infants who suffer from a "pre-natal produced neurosis" which makes their adjustment to life outside the mother's body difficult. Research studies of animals find that excessive maternal stress can curb masculinity in boys.

Prolonged and extreme maternal stress during the period of the fetus frequently causes more illness during he first three years of the child's had a more favourable fetal environment. Children whose mothers were under great stress during pregnancy also show more "free floating anxiety", although they can still perform their daily routines. Such anxiety has an adverse effect on their ability to learn to remember, and to reason to their full capacities. As a result, they seem to be less bright than they are actually one.

Unquestionably, one of the most serious effects of maternal stress during pregnancy is on children's post-natal adjustment to family members. Because of their hyperactivity, excessive crying, and other indications of poor adjustment to post-natal life, they are regarded as "difficult" babies. Attitude of family members towards them are then far less favourable than they would have been had they made better adjustments to postnatal life.

As they grow older, they sense these unfavourable attitudes on the part of family members and later on, the part of peers, teachers and other outsiders. Feeling unloved and rejected, they often show below average physical development, hyperactivity, and lag in developing motor skills and speech, and learning problems. All these lead to poor personal and social adjustments. In 13-16 per cent of all young children, psychological adaptation is hampered by mild or severe neurodevelopment disorders that range from hyperactivity, learning disabilities, language delay and motor abnormalities to autistic spectrum disorders and cerebral palsy.

Animal studies have shown that the prenatal period should not be neglected when one is interested in early determinants of development (Weinstock, 1997). As a result there is now an increasing recognition of the role played by rental factors in the development of subsequent neuropsychiatry impairment, particularly in term both infants.

EFFECTS OF PRENATAL MATERNAL STRESS

Pregnancy is a special time for a woman and her family. It is a time of many changes—in a pregnant woman's body, in her emotions and in the life of her family. As welcome as they may be, these changes often add new stresses to the lives of busy pregnant women who already face many demands at home and at work.

Stress, however, does not have to be all bad. When managed properly, stress can provide us with the drive to meet new challenges. A pregnant woman (or anyone else) who feels she is coping well with stress taking good care of herself, feeling energized, rather than drained, and functioning well at home and work probably does not face health risks from stress.

However, when physical or emotional stress builds up to uncomfortable levels, it can be harmful for pregnant women. In the short term, a high level of stress can cause fatigue, sleeplessness, anxiety, poor appetite or overeating, headaches and backaches. When a high level of stress continues for a long period, it may contribute to potentially serious health problems, such as lowered resistance to infectious diseases, high blood pressure and heart disease. Studies also suggest that high levels of stress may pose special risks during pregnancy.

The good news is that stress is unlikely to cause problems in pregnancy for most women. Pregnant women who are concerned about the level of stress in their lives, and their ability to cope with it, should talk with their health care provider. A health care provider may be able to point a woman to resources in her community and show her simple, effective steps to help her reduce and cope with physical and emotional stress.

Babies who are born prematurely are often low birth weight. However, studies suggest that babies of women who suffer from high levels of stress and anxiety are more likely to be born low birth weight even when born at full term. Some stress-related hormones (such as nor epinephrine) may constrict blood flow to the placenta, so the baby may not receive the nutrients and oxygen it needs for optimal growth.

Stress also may exert potential adverse effects indirectly by affecting the pregnant woman's behaviour. For example, women who experience high levels of stress may not follow good health habits. They may skip meals or not choose nutritious foods, or they may react to stress by reaching for cigarettes, alcohol or illicit drugs, all of which have been linked to low birth weight. Use of alcohol and certain illicit drugs also increases the risk of birth defects.

Studies also suggest that high levels of stress may contribute to other pregnancy complications. A few studies suggest that women with stressful jobs may be slightly more likely than women with low-stress jobs to develop pre-eclampsia (a pregnancy-related disorder that includes high blood pressure and can result in poor fetal growth and other problems). There also is a concern that high levels of stress may increase the risk of miscarriage.

REASONS FOR PRENATAL MATERNAL STRESS

In women cerion life conditions, such as, Menstruation, gestation, puerperium, menopause and climetarium are producers of stress whose effects can be manifested in physic laboral areas.

Pregnancy-related discomforts (such as nausea, fatigue, frequent urination, swelling and backache) can be stressful, especially if a pregnant woman attempts to accomplish everything she did prior to pregnancy. A pregnant woman can help reduce her stress by recognizing that these symptoms are temporary, and her health care provider can recommend ways to cope with them. She also can consider cutting back on unnecessary activities when she is uncomfortable.

Hormonal changes may be partly responsible for the mood swings experienced during pregnancy. These mood swings are common and normal, so a pregnant woman should not be overly concerned about them. However, she should keep in mind that mood swings may make it more difficult for her to cope with stress.

In addition, many pregnant women and their partners worry about the health of their unborn baby, their ability to cope with labour and delivery and their ability to become good parents. Added financial responsibilities are another common source of stress, especially if the parents anticipate a reduction in income—whether brief or long-term—after the baby is born. All of these worries can be magnified if the pregnancy is high-risk, in which the pregnant woman must significantly reduce her activity and, possibly, leave her job early.

Stress also may exert potential adverse effects indirectly by affecting the pregnant woman's behaviour. For example, women who experience high levels of stress may not follow good health habits. They may skip meals or not choose nutritious foods, or they may react to stress by reaching for cigarettes, alcohol or illicit drugs, all of which have been linked to low birth weight. Use of alcohol and certain illicit drugs also increases the risk of birth defects.

Studies also suggest that high levels of stress may contribute to other pregnancy complications. A few studies suggest that women with stressful jobs may be slightly more likely than women with low-stress jobs to develop pre-eclampsia (a pregnancy-related disorder that includes high blood pressure and can result in poor fetal growth and other problems). There also is a concern that high levels of stress may increase the risk of miscarriage. Studies to date suggest that these risks—if they exist at all—are likely to be small for most women.

WAYS TO REDUCE STRESS DURING PREGNANCY

Each pregnant woman needs to identify the personal and work-related sources of stress in her life and develop effective ways to deal with them. Any woman, whether pregnant or not, can cope better with the stresses in her life if she is healthy and fit.

A pregnant woman should eat a healthy diet, get plenty of sleep. It is important that a pregnant woman not skip meals because the fetus needs constant nutrition, and long fasts put stress on a woman's body. She should eat three meals plus at least two healthy snacks, or five to six small meals a day. Exercise keeps pregnant women fit, helps prevent some common discomforts of pregnancy (such as backache, fatigue and constipation) and relieves stress.

Having a good support network which can include the pregnant woman's partner, extended family, friends and others also helps a pregnant woman relieve stress. These individuals may provide information, emotional support or help with tasks around the home. Some studies suggest that having good support actually may reduce the risk of preterm labour and low birth weight, especially for poor, high risk women or women who are feeling stressed.

A number of stress reduction techniques have been used successfully in pregnancy. These include biofeedback, meditation, guided mental imagery and yoga (specifically for pregnant women). Unless a pregnant woman has practiced these techniques previously, she may need instruction from an expert. Childbirth education classes teach relaxation techniques and help reduce anxiety by educating parents-to-be-about what to expect during labor and delivery.

McCubbin has developed a simple relaxation plan that any pregnant woman can use:

- Relax for the health of your baby and yourself. Maternal stress can affect your developing baby.

- Allow sufficient time to relax each day. Relaxation is important for you and your baby.
- Get comfortable. A quiet room with no phone works best. Lying down or reclining is good. Lie slightly tilted to one side with your belly (and baby) partially supported by a pillow.
- Prepare mentally. Clear your mind of distractions and focus on your relaxation.
- Take control. You control the relaxation you give to your body and your baby.
- Focus on your breathing. Use slow, steady, deep breaths from your belly, not your chest.
- Monitor your muscles. Learn to recognize tension in your body's major muscle groups.
- Release the tension in each muscle group. Become familiar with the feel of tension dissolving.
- Imagine yourself in your favourite restful place—maybe on the beach, by a stream or on a mountain top.
- Practice and enjoy the pleasant feelings that you have given to yourself and your baby. Do it at least once a day for 20 to 30 minutes. Relax throughout pregnancy.

Fortunately, most women adjust well to the physical and psychological changes of pregnancy. However, if a pregnant woman feels overwhelmed by stress, she always should consult her maternity health care provider who may recommend that she see a mental health professional.

NEED FOR THE STUDY

Research demonstrates that chronic prenatal maternal stress contributes to poor birth outcome, particularly low birth weight and preterm delivery. It has been suggested that prenatal health behaviour mediate the relationship between stress and birth outcome. In particular proper nutritional practices are critical to the health of mother and baby. Stress has been shown to have adverse effects on the eating behaviour of pregnant women. Women who experience greater stress during pregnancy were less likely to report eating a nutritional diet, taking vitamins, and drinking adequate fluids, particularly when they experienced interpersonal stress or stress evoked by fear of physical pain. Health relevant consequences of maternal obesity and poor nutrition in pregnancy as a response to stress highlight the importance of encouraging stress management for pregnant women.

Prenatal stresses are negatively related to infant's mental and motor development. In addition, woman who experienced many fears during pregnancy have a risk of getting and with adaptation problems and difficult behaviour in first month of life.

OBJECTIVE OF THE STUDY

The objectives of the present study are the following:

1. To study the stress level of pregnant women belonging to different age groups.
2. To study the stress level of the first, second and third pregnancy in a pregnant women.
3. To study the stress level of pregnant women in different stages of pregnancy.
4. To study the stress level of working and non working pregnant women.
5. To study the stress level of pregnant women having different levels of educational qualification.
6. To study the stress level of pregnant women belonging to joint or nuclear families.
7. To study the stress level of pregnant women belonging to different socio-economic status.

LIMITATIONS OF THE STUDY

This study was restricted to very few private clinics and hospitals. In order to make studies in this field more comprehensive, studies have to be replicated in other hospitals where pregnant belonging to various categories do get admitted. Due to time constraint only a sample of 50 pregnant women was taken. The study would have been more elaborate if more number of pregnant women were included in it. For Illiterate pregnant women the investigator had to read about the questionnaire and note down their responses.

This study would be more useful for Gynecologists, Researchers, Psychiatrists and Counselors. If we could do the stress reduction counseling to the pregnant women it is very useful to their health and baby, which was not possible with limitations of time and samples.

2

REVIEW OF RELATED RESEARCH

The survey of related literature and studies also helps to avoid the risk of duplication. It helps the investigator to see whether the evidence already available solves the problem adequately without further investigation and thus to avoid the risk of duplication. Every investigator is expected to know what sources are available in his field of enquiry, which of them he is likely to use and where and how to find them.

Research takes the advantage of the knowledge which has accumulated in the past as a result of constant human endeavour. It can never be undertaken in isolation of the work that has already been done by a research. A careful review of the research journals, books, dissertations, theses and other sources of information on the problem to be investigated is one of the important steps in the planning of any research study.

Review of related literature, besides to allow the researcher to acquaint herself with current knowledge in the field or area in which he is going to conduct his research, serves the following purposes: 1. The review of related literature enables the researcher to define the limits of his field. It helps the researcher to delimit and define his problem; 2. By reviewing the related literature, the researcher can avoid unfruitful and useless problem areas. He can select those areas in which positive findings are very likely to result and his endeavours would be likely to add to the knowledge in a meaningful way;

3. Through the review of related literature, the researcher can avoid unintentional duplication of well established findings. It is no use to replicate a study when the stability and validity of its results have been clearly established; 4. The final and important specific reason for reviewing the related literature is to know about the recommendations of previous researchers for further research which they have listed in their studies.

The first step in reviewing the related literature is the identification of the material that is to be read and evaluated. The identification can be made through the use of primary and secondary sources available in the library or elsewhere.

In primary sources of information, the author reports his own work directly in the form of research articles, books, monographs, dissertations or theses. Primary sources give the researcher a basis on which he can make an own judgment of the study. Though consulting such sources is a time consuming process for a researcher, yet they provide a good source of information on the research methods used.

In secondary sources, the author compiles and summarizes the findings of the work done by others and gives interpretation of these findings. The author usually attempts to cover all the important studies in areas in the form of encyclopaedia indexes, abstracts, bibliographies, references and quotation sources. Working with secondary sources is not time-consuming because of the amount of reading required. The disadvantage of the secondary sources, however, is that the reader is depending upon someone else's judgment about the important and significant aspects of the study.

The decision concerning the use of primary or secondary sources depends largely on the nature of the research study proposed by the researcher. If it is a study in an area in which much research has been reported, then a review of the primary sources would be a logical first step. On the other hand, if the study is in an area in which little or no research has been conducted, a check of the secondary sources is more logical. Sources of information, whether primary or secondary, are found in the library. The researcher must, therefore, develop the expertise to use resources without much loss of time and energy. To aid the researcher in locating, selecting and utilizing the resources, study guides are provided in relation to their use in educational research.

Citing studies that show substantial agreement and those that seem to present conflicting conclusions helps to sharpen and define understanding of existing knowledge in the problem area, provides a background for the research project and makes the reader aware of the status of the issue. Parading a long list of annotated studies relating to the problem is ineffective and inappropriate. Only those studies that are plainly relevant, competently executed and clearly reported should be included in the review of related literature. (Bhaskara Rao, 1997).

The following are the studies found relevant to some extent to the present study.

Williamson and Fam Pract (1981) studied about the stress and coping in first pregnancy. This study examined the role of the family physician in caring for nine couples during first pregnancy through the post-partum period. His study was evaluate their support, stresses and coping styles were included when it is recognized, and stresses were reduced provision of information, discussion and reassurance.

Georgas and Giakoumaki (1984) studied about the relationship of psychosocial stress to symptoms during pregnancy, obstetric complications, family planning and breast feeding. The level of psychosocial stress experienced during the 6 month period prior to delivery was positively associated with the occurrence of obstetric complications. The level of stress experienced more than 6 months prior to delivery was not, the most common stresogenic events reported by women who had obstetric complications were: 1. High anxiety about the health status of the fetus; 2. death of a loved one; 3. arguments with parents or spouse. The level of psychosocial stress was also positively related to occurrence of explained by differences in psychosocial stress levels.

Kemp (1986) Studied about the psychosocial impact of a high—risk pregnancy on and planning care for the family is described. The model depicts four major concepts that the nurse must consider in determining how the family is integrating, interpreting and adapting to the high risk pregnancy, health status of the pregnancy, support available to the family, and the family adaptation to a high risk pregnancy are discussed.

Arizmendi and Affonso (1987) studied about stressful events related to pregnancy and post-partum. An 221 women were used to identify these stressors. These are: First Trimester women (N=81), Third Trimester women (N=80), and post-partum women (N=60). Stressors were rated according to their frequency and intensity. The results focused both on the most intense stressors for each group as well as the changes in stressor intensity across the groups.

Levin Jeffery (1988) researched on maternal stress and pregnancy outcomes. Research findings were reviewed with pregnancy outcomes grouped into four categories; in low birth weight, prematurity, antepartum and intrapartum complications. In general, literature reveals that anxiety is scattered evidence linking stress to low birth weight.

Mercer, *et al* (1989) studied the effects of stress on family functioning during pregnancy. Family functioning was tested in four groups of expected parents (age 18-61 years). The groups included 153 hospitalized women, 208 low risk women and 147 of their partners. Both partners in the high risk situation reported greater discrepancy in family functioning than in partners in low risk situation.

Hoffman and Perinatal (1990) studied about the stress factors related to antenatal testing during high risk pregnancy. This testing brings with it emotional stress as well as the ability to diagnose abnormalities. Care of the pregnant patient must include identifying stressors and designing strategies to reduce them. These interventions will make procedures as easy as possible for the antenatal patient and staff.

Murray, *et al* (1990) did a screening for depression pregnancy using Edinburgh depression scale. 100 women were studied between 28-34 weeks attending antenatal clinic. The scale was able to identity the pregnant women with RDC (Research Diagnostic Criteria) with major depression but was less effective in detecting those with RDC minor depression.

Sammons (1990) studied about the psychosocial aspects of second pregnancy. These experiences are described including maternal tasks and sources of anxiety for second gravidas, as a basis for improving care to expanding families. Clinical implications are suggested that meet the special needs of second time mothers regarding time limitations to meet multiple demands, family relationship changes and concerns about maternal health and fatigue.

Moore, *et al* (1991) did a comparison of emotional state and support in women of high state and low risk for preterm birth with diabetes in pregnancy and in non-pregnant professional women. Psychological profiles were developed for 48 pregnant women at high risk for preterm labour (PTL), 25 at low risk for PTL, 73 pregnant diabetic women and 16 non-pregnant professional women of childbearing age. Instruments included Depression adjective checklist, the perceived stress scale, the Eysenck personality inventory and measure the state anxiety. Result suggests the socio-economic status may be as important as medical risk as a source of distress among pregnant women.

Goldenberg, *et al* (1991) studied the maternal psychological characteristics and interactive growth retardation. A sample of 184 multiparous indignant women on 6 psychosocial questionnaires was compared to the risk of delivering in infant with intrauterine growth retardation (IUGR). In the 2nd trimester, scales for stress, anxiety, social support, and self-esteem had been compared. The result were addictive in that the poorer the score, higher the Rate of IUGR.

Pond, *et al* (1992) studied about a comparison between adolescent and adult women on prenatal anxiety and self-confidence. This study was conducted on anxiety and self-confidence during pregnancy in adolescents and adults. A significant negative correlation was found for the total group between state and Trait Anxiety and self-confidence. The higher were found for anxiety level, the lower the self-confidence. No confidence in this population. The finding of a significant negative relationship between prenatal anxiety and boast self-confidence has implications for nursing practice and research. Assessing the needs, concerns, and self-confidence of each pregnant woman through the perinatal period will allow the planning of appropriate women interventions. Individual or group, education and counselling may help alleviate anxiety and boast self confidence.

Fam Pract and Forde (1992) studied about the pregnant women's ailments and psychosocial conditions. It may cause anxiety and reduce the quality of life. A higher number of ailments were reported by women with psychosocial problems and heavy physical work. Abdominal pain was more often reported by women assessed as having psychosocial problems. Based on background knowledge, clinical observations and a some what extended history on the psychosocial situation, it is possible to identity a subgroup of women who experience a more than 50 per cent increase in the frequency of such ailments.

Libeberman, *et al* (1993) studied the anxiety during pregnancy at the time of the Gulf War. Comparisons of anxiety of Israeli women with normal pregnancy and those with at high-risk pregnancy. Interviewed 81 pregnant Israeli women (19-42 years) at the time of gulf war. 53 was at risk pregnant and 28 was normal pregnant. It was suggested that, the already stressed at risk group shut out the additional anxiety generated by the dangers of the war.

Sjogren (1997) studied the reason for anxiety about childbirth in 100 pregnant women. A sample of N-100 referred to antenatal center, because of extreme fear of childbirth, were interviewed and it was found that anxieties over childbirth is related to fundamental human feelings, lack of trust fear of female incompetence and fear of death, fear of pain is important but not predominant.

Coasta, *et al* (1998) studied the impact of psychological and lifestyle variables basis beginning in the third months of pregnancy. Measures of daily stress state anxiety and pregnancy specific stress were taken monthly. The result indicates that certain psychological and lifestyle variables may be differently associated with complications occurring at various phases of pregnancy.

Tam (1998) Studied about his psychological aspects of pregnancy in the military. His study examine the past 20 years, more women have joined the armed forces than ever before. Three types of psychological stressors are identified lack of social supports, the pressures of minority status, and the institutional reactions to gender roles. The author proposes mandatory education for military leaders regarding sensitivity to women's reproductive issues as well as specialized briefings for women inductees on the realities and risks of pregnancy in the military.

Dragonas (1998) studied on early childhood education. He examined the psychological and psychosocial aspects of prenatal care. The effect provision of prenatal care has on women as well as their role in their own care is raised. The need is stressed for further sophisticated longitudinal and multivariate research exploring further the causative links between quality of prenatal care, pregnancy outcome, and subsequent child development.

Haglund (1998) studied the perinatal assessment of psychosocial risk. The reviews several major psychosocial risk factors together with instruments that have been utilized to assess them during the perinatal period. Formal constructs reviewed include anxiety, depression, self concept, general attitudes, life events, stress, adaptation, social support, marital and family functioning, and the home environment.

CMAJ (1999) studied about the physical abuse during pregnancy: Prevalence and risk factors. Violence during pregnancy is a health and social problem that poses particular risks to the women and her fetus. Perceived stress and number of negative life events in the preceding year were also predictors of abuse. Physical abuse affects a significant minority of pregnant women and is associated with stress lack of perceived support and a partner with a drinking problem.

Klock, *et al* (2000) studied about the psychosocial status of In Vitro Fertilization (IVF) patients during pregnancy. This study determines whether women who get pregnant who conceived 74 women who became pregnant via IVF and 40 women conceiving without medical intervention. Pregnant IVF women are similar psychologically to women who become pregnant naturally on dimensions of self-esteem, depression, and anxiety at 12 and 28 weeks gestational age. The IVGF group, not controls, reported improved self-esteem and decreased anxiety as the pregnancy progressed.

Sammons, L.N. (1990) studied about the psychosocial aspects of second pregnancy. The second pregnancy experiences are described including maternal tasks and sources of anxiety for second gravidas, as a basis for improving care to expanding families. Clinical implications are suggested that meet the special needs of second time mothers regarding time limitations to meet multiple demands, family relationship changes and concerns about maternal health and fatigue. Strategies for enhancing care involve using childbirth and sibling classes, modifying health care and information from primary care providers, mobilizing supportive services and resources, and influencing policies to meet maternal and family needs.

Jean (2002) conducted a research to determine the effects on puppies of the prenatal stress and of the environment of the pregnant females. The study investigates whether a mild or a hard maternal stress could influence the performance (learning ability and emotional stability) of the puppies. That impact has measured on a variety of developmental and temperamental measures.

Nancy Dole (2003) studied stress in pregnancy tied to premature delivery. Pregnant women who said they were experiencing high levels of stress from events such as, marital separation, problems with in-law or issues at work, were 80 per cent more likely to have a preterm delivery than those who reported low stress levels. Overall there were 71 preterm births and 402 full term births among women reporting high stress, compared to 39 preterm and 416 full term births in the low stress group.

Austin (2003) studied the psychosocial assessment and management of depression and anxiety in pregnancy. This article aims to provide an overview of psychosocial assessment and the detection and management of depression and anxiety disorders in pregnancy. Psychosocial assessment of all pregnant women is an integral part of good antenatal care.

Eberhard (2005) studied the mood disorders during pregnancy. It may have a negative affect on self care and pregnancy outcome that affects the mother directly and the child indirectly. The knowledge about long-term neurobehavioural effects in the offspring is still limited for all agents and requires further investigation. Possible adverse effects of fetal exposure must be balanced against the adverse effects of an untreated maternal mood disorder.

Flynn, *et al* (2006) studied about depression treatment among pregnant women. His study was to provide information on rates of depression treatment among pregnant women at risk for depression and among those with clinician diagnosed current major depressive disorder (MDD) and to examine predictors of depression treatment. Among women with a current MDD Diagnosis, most of whom were experiencing a recurrence, 335 were currently receiving any depression treatment. The presence of current MDD was not found to be related to use of treatment.

O'Brien (2007) studied depression, anxiety, irritability, and stress in pregnancy. He determine the effectiveness of maintaining antidepressants during pregnancy, as measured by changes in symptoms of depression, anxiety, irritability and stress following reassuring evidence based counselling. Out of the 58 women who enrolled in the study, 38 completed 75 per cent of the follow ups. Eight women (14%) discontinued their medication during the study. Depression scores were highest at enrollment in the first trimester and decreased as pregnancy progressed. When data from all women, regardless of dose adjustments, were analyzed no statistically significant differences were seen the cut-off for depression throughout the study period. Irritability, anxiety, and stress scores were not found to be statistically different at any time point during the study.

3

METHODOLOGY OF RESEARCH

Planning is a necessary step for a good research. Design is the heart of any research. In this chapter, the aspects that are concerned with the design of the present study have been discussed in detail. Research procedure followed includes the operational definitions of the different terms used, the various hypotheses that were framed for verification and the rationale of these hypotheses. Selection of the sample includes the sampling techniques used, the reasons for selection of a particular sampling technique and the selection of sample according to different variables. Selection of tool covers the description of tool selected, testing of its suitability for the present study, and procedure followed in the administration of the tool to collect the data required for this study.

The present study was conducted to examine the prenatal stress among pregnant women. The study was intentionally done on pregnant women since the follow up of clients to the hospital is encouraging. Earlier studies have indicated that the pregnant women had stress and anxiety about their pregnancy and prenatal development of child.

HYPOTHESES OF THE STUDY

The following are the hypotheses formulated for the present study:

1. There would not be any significant difference among the stress levels of pregnant women belonging to different age groups.

(a) There would not be any significant difference between the stress levels of pregnant women of below 25 years and 25 years of age.

(b) There would not be any significant difference between the stress levels of pregnant women between 25 years and 30 years of age.

(c) There would not be any significant difference between the stress levels of pregnant women of 30 years and above 30 years of age.

2. There would not be any significant difference among the stress levels of pregnant women on the basis of number of pregnancies.

(a) There would not be any significant difference between the stress level of pregnant women having 1st and 2nd pregnancies.

(b) There would not be any significant difference between the stress level of pregnant women having 2nd and 3rd pregnancies.

(c) There would not be any significant difference between the stress level of pregnant women having 1st stand 3rd pregnancies.

3. There would not be any significant difference among the stress levels of pregnant women irrespective of their months of pregnancy.

(a) There will not be any significant difference between the stress levels of 5 months and 6-7 months pregnant women.

(b) There would not be any significant difference between the stress levels of 6-7 months and 8-9 months pregnant women.

(c) There would not be any significant difference between the stress levels of 5 months and 8-9 months pregnant women.

4. There would not be any significant difference between the stress levels of Working and Non-working pregnant women.

5. There would be not be any significant difference among the levels of pregnant women having different level of education.

   (a) There would not be any significant difference between the stress levels of pregnant women having school education and no education.

   (b) There would be any significant difference between the stress levels of pregnant women having UG level of education and no education.

   (c) There would not be any significant difference between the stress levels of pregnant women having PG level education and uneducated.

   (d) There would be significant difference between the stress levels of pregnant women having school and UG level of education.

   (e) There would not be any significant difference between the stress levels of pregnant women having school and PG level of education.

   (f) There would not be any significant difference between the stress levels of pregnant women having UG and PG level of education.

6. There would not be any significant difference between the stress levels of pregnant women belonging to joint and nuclear families.

7. There would not be any significant difference among the stress levels of pregnant women respect of their socio-economic status.

SAMPLE OF THE STUDY

The study was conducted at the Southern Railway Hospital, Perambur, Chennai. This Centre renders medical and health care services to out and in patients of different types health problems. In this hospital, the clinical laboratory for diagnosing pregnancy conformation, counseling and HIV testing facilities are provided on weekdays only. About 5 to 10 pregnant women visit the hospital daily.

The study was conducted on a sample of 50 pregnant women at the Southern Railway Hospital, Perambur, Chennai. The sample included those who had undergone pregnancy testing and were confirmed and those who referred to counselling and HIV testing. The investigator approached Chief Medical Officer and explained about the study and got permission. Then the investigator approached these pregnant women and requested them to participate in this study. The patients were made to feel at ease and an initial rapport was established. The questionnaire was given to pregnant women and they were asked to fill them as per the instructions.

The total sample of the study consisted of 50 pregnant women. Among them 16 were below twenty-five years, 22 were between 25 to 30 years of age and 12 were more than thirty years. The mean age of the total patients is 27.79 years.

TOOL OF THE STUDY

Keeping in view the objectives of the study and the nature of the research, questionnaire seems to be an ideal choice. Stress Assessment Questionnaire was selected to suit the specific needs of the study as well as the sample to be investigated. As the sample is from Southern Railway Hospital, Perambur, Chennai, wherein the vernacular language Tamil is mostly used, it was felt that the questionnaire have to be translated into Tamil language.

The questionnaire begins with instructions to the patients regarding the nature and purpose of the research. Necessary instructions were given to fill each item. The general information regarding the pregnant women, namely, their name, age, occupation, education and type of family and socio-economic status was obtained.

Stress Assessment Questionnaire was used in the present study. The details of the questionnaire are provided below.

Stress Assessment Questionnaire (SAQ)

The Stress Assessment Questionnaire (SAQ) was developed by Allen Cameron's aims to help people recognize and manage stress in their daily lives. The SAQ assesses seventeen dimensions of stress covering possible sources, common symptoms, coping style, personality factors and the impact on mental health. The SAQ is designed for the counselee to understand the things that may be coping stress in their life and how they respond to pressure and stress. There are 128 questions and it should take her about 30-35 minutes to complete.

What SAQ Covers?

Key Area	Dimension	Description
Source of Stress	1. Work	Dissatisfied with job, Stress arises from various work conditions.
	2. Relationship	Suffers stress as a result of relationship difficulties in personal life.
	3. Parenting	Experiences stress as a result of overload from childbearing and family.
	4. Incidents	Experiences stress as a result of critical incidents.
Symptoms of Stress	1. Emotional	Worries, feels afraid, has intrusive thoughts, may have panic attacks.
	2. Behavioural	Feels tired, works less efficiently, has difficulty in concentrating.
	3. Physical	Multiple symptoms such as indigestion, headaches, skin complaints, etc.
Coping Style	1. Social support	Talks through problems, seeks social support to help reduce stress.

(Contd...)

Key Area	Dimension	Description
	2. Self-Regulation	Recognizes and manages feelings and emotions.
	3. Problem solving	Seeks to understand, control and improve situation.
	4. Distraction	Seeks distractions to reduce pressure and take mind off things.
	5. Health	Keeps in shape, eats healthy diet, moderates drinking.
Personality Factors	1. Procrastination	Put things off, avoids completing projects, lacks initiative.
	2. Perfectionism	Holds unrealistic standards about self and other people.
Mental Health	1. Self-esteem	Lacks self-respect, feels worthless, judges self to be a failure.
	2. Depression	Feels unhappy, guilty and inadequate, loses motivation and interest.
	3. Anxiety	Worries excessively, has feeling of fear, relives upsetting events.

Psychometric Properties

The table below summarizes the psychometric properties of the Questionnaire.

Designed to measure	:	16 dimensions of stress covering sources of stress, common symptoms, coping style and personality factors.
Questionnaire Type	:	Normative
Design Model	:	Conceptual/deductive approach
Number of Questions	:	169
Number of scales	:	16 scales of 8 items, positively and negatively keyed.

(Contd...)

Response Format	:	5 point Likert Scale
Completion Time	:	30-35 minutes
Norms	:	1000 professional and managerial balanced equally between males and females.
Reliability	:	Median consistency of scale is 0.8. Stress factor pairs share less than 50 per cent common reliable variance.
Validity	:	Median correction with IPIP marker variable is 0.82.Principal components FA confirmed presence of these factors measuring stress symptoms, coping style and personality dimensions. Correlation between stress symptoms, factors and the Holms-Rahe Social Adjustment Scale is 0.29.
Report Format	:	Interpretative text and profile chart finished to publication quality output.

Scoring

The Stress Assessment scale is a self-report measure whose items are scored on a five point Likert scale that ranges from Hardly ever (scored 1), Seldom (scored 2), Sometimes (scored 3), Often (scored 4) and Nearly always (Scored 5). A pregnant woman completing the measure is instructed to indicate how they feel about pregnancy and its effects on their life. All items scored and there is no reverse scoring. Raw scores on the 128 items are summed to produce a total score that ranges from 128 to 640. A high score indicates low level of stress to pregnancy.

The scale consists of the 128 items and five key areas with seventeen dimensions. The five key areas and seventeen dimensions are the following.

DATA ANALYSIS

The data obtained from the sample of 50 pregnant women was scored and analyzed. The analysis involved Mean, Standard Deviation and the application of t-tests, analysis of variance. The results obtained from the analysis are presented in the next chapter.

4

RESULTS AND DISCUSSION

The present study was conducted to assess the stress level among pregnant women. This chapter mainly deals with different types of statistical analysis like descriptive and inferential methods. Descriptive methods like mean and standard deviation have been calculated to analyze the data in a meaningful way whereas the inferential statistical techniques like analysis of variance and 't' test have been used to test the formulated hypotheses. The results of the present study are presented and discussed after a presentation of a sample profile.

SAMPLE PROFILE

The profile of the sample is provided in Table 4.1. It can be observed that the sample consists of 50 pregnant women. The sample included that age, number of pregnancies, moths of pregnancy, occupation, education, type of family, socio-economic status of pregnant women. The details of sample are provided in Table 4.1.

The subsequent results of the thesis are divided into 7 sections dealing with age groups, number of pregnancies, moths of pregnancy, occupation, education, type of family and socio-economic status of pregnant women.

Table—4.1

Profile of the Sample, the Pregnant Women

Variable	Sub variable	Sample
Age	1. Below 25 years	16
	2. Between 25 to 30 years	22
	3. Above 30 years	12
Number of pregnancies	1. First pregnancy	27
	2. Second pregnancy	16
	3. Third pregnancy	7
Months of pregnancies	1. Upto 5 moths	10
	2. 6-7 months	22
	3. 8-9 months	18
Occupation	1. Working	22
	2. Non-working	28
Education	1. Post-graduate	7
	2. Graduate	13
	3. School	24
	4. Uneducated	6
Type of family	1. Joint family	29
	2. Nuclear family	21
Socio-economic status	1. Lower income	25
	2. Middle income	14
	3. Higher income	11

1. Age Groups

The relationship between age and stress is described in Table 4.2.

The Table 4.2 indicates that there is a significant difference in the level of stress among pregnant women irrespective of their age. In the age group of below 25 years, the reason may be their lack of mental maturity and fear of labour pain. In the second age group of 25 to 30 years of age, though they are physically and mentally mature the fears and in some cases lack of social support can increase their stress in pregnancy. In the above 30 years of age group, the chances of complicated pregnancy are more compare to younger age group.

Table—4.2

The Stress Level Among the Pregnant Women at Different Age Groups

Sl. No.	Group	Mean	SD	F-value
1.	Up to 25 years	400.18	39.42	
2.	Between 25 and 30 years	387.31	37.21	1.85*
3.	Above 30 years	415.83	50.97	

* Significant at 0.01 level.

2. Number of Pregnancy

The relationship between the number of pregnancy and stress is explained in Table 4.3.

Table—4.3

The Stress Level Among First, Second and Third Time Pregnant Women

Sl. No.	Group	Mean	SD	F-value/ t-value
1.	1st pregnancy 2nd pregnancy 3rd pregnancy	389.40 393.68 443.00	40.70 39.70 25.86	5.45*
2.	1st pregnancy 2nd pregnancy	389.40 393.68	40.70 39.70	0.33**
3.	2nd pregnancy 3rd pregnancy	393.68 443.00	39.70 25.86	3.53*
4.	1st pregnancy 3rd pregnancy	389.40 443.00	40.70 25.86	4.27*

** Not significant at 0.01 level.

* Significant at 0.01 level.

The Table 4.3 indicates that there is a significant difference in the level of stress among the pregnant women irrespective of number of pregnancies. This finding has rejected the major null hypothesis number I. Further, to investigate the significance level of stress among the pregnant women those who have more number of pregnancies, the 't' test has been computed to test the sub-hypotheses formulated to this process.

The above Table, which is not significant at any level, indicates that there is no significant difference in the level of stress between the pregnant women having 1st and 2nd pregnancies. This result accepts the hypothesis 2(a), which is formulated for this purpose. The reason for the stress in the 1st pregnancy may be due to the anxiety for the condition 'pregnancy' and lack of knowledge and experience about it. Same way, in 2nd pregnancy, though they know the process yet the syndrome occurring during the pregnancy period and the outcome of labour make them develop stress.

The above table indicates that there is significant difference in the level of stress between the women having 2nd and 3rd pregnancies. The result rejects the null hypothesis 2 (b), which is formulated for this purpose. In this table the 'mean' shows that the women having 2nd pregnancy are also having much stress but 3rd time pregnant women are more stressed. The reason could be that nowadays the third baby increases the pressure on the women's health as well as on the family's finances.

The above table indicates that there is significant difference in the level of stress between the women having 1st and 3rd pregnancies. The result rejects the null hypothesis 2 (c), which is formulated for this purpose. The reason could be that in first pregnancy, despite all the stressful factors, enthusiasm is there for the first baby. But in the third pregnancy it somehow feels like a burden which she has to carry, in most of the cases, either unwanted pregnancy or for a 'boy' baby.

3. Pregnancy in Months

The relationship between pregnancy in months and stress is described in Table 4.4.

Table—4.4

The Stress Level Among the Pregnant Women at Different Stages of Pregnancy

Sl. No.	Group	Mean	SD	F-value
1.	Up to 5 months	409.9	44.10	
2.	Between 6 and 7 months	390.18	43.02	0.83*
3.	Between 8 and 9 months	401.72	40.57	

* Not significant at 0.01 level.

The Table 4.4, Indicates that there is no significant difference in the level of stress among the pregnant women irrespective of their period of pregnancy. This finding has accepted the major null hypothesis number III. These results indicate that there is no need for further investigation of the significance level of stress among the pregnant women at different stages of education. That is why, the 't' test has not been computed to test sub-hypothesis formulated for this purpose. The reason may be the same fears, anxieties and risk involved in pregnancy. In all stages of pregnancy, common syndromes can occur, i.e., pain in lower back, cramps, oedema, insomnia and nightmares, varicose veins etc.

4. Working And Non-working

The relationship between working and non-working and stress is explained in Table 4.5.

Table—4.5

The Stress Level Among the Working and Non-working Pregnant Women

Sl. No.	Group	Mean	SD	t-value
1.	Working	407.86	35.31	1.48+
2.	Non-working	390.18	46.18	

* Not significant at 0.01 level.

The Table 4.5 indicates that there is no significant difference in the level of stress between working and non-working pregnant women. This finding has accepted the major null hypothesis 4. Working Pregnant women have to cope up with the work pressure as well as the physical change during the pregnancy. Non-working pregnant women do not have that extra income and syndromes occurring during the pregnancy are same for them. So the stress level is same in pregnant women irrespective of their job status.

5. Education

The relationship between education and stress are observed in Table 4.6.

Table—4.6

The Stress Level Among the Pregnant Women Different Level of Education

Sl. No.	Group	Mean	SD	F-value/ t-value
1.	Uneducated	430.50	39.95	2.85*
	School	404.29	25.66	
	Graduate	376.46	52.05	
	Post-graduate	390.57	54.81	
2.	Uneducated	430.50	39.95	1.52+
	School	404.00	25.66	
3.	Uneducated	430.50	39.95	2.48*
	Graduate	376.46	52.05	
4.	Uneducated	430.50	39.95	1.51+
	Postgraduate	390.57	54.81	
5.	School	404.29	25.66	1.81*
	Graduate	376.46	52.05	
6.	School	404.29	25.66	0.64+
	Post-graduate	390.57	54.81	
7.	Graduate	376.46	52.05	0.55+
	Post-graduate	390.57	54.81	

\+ Not significant at 0.01 level.

* Significant at 0.01 level.

This finding has rejected the major null hypothesis number 5. Further, to investigate the significance level of stress among the pregnant women having different level of education, the 't' test has been computed to test sub-hypothesis formulated for this purpose.

The Table 4.6, which is not significant at any level indicates that there is no significant difference in the level of stress between the pregnant women who are uneducated and those having level of education.

This test accepts the null hypothesis 5(a), which is formulated for this purpose. The reason could be that the school level of knowledge does not include the facts like pregnancy and delivery. So they are also like uneducated in these matters. What knowledge they get from elders and friends, that is more confusing and distressing than helping.

The above table indicates that there is significant difference in the level of stress between uneducated and graduate pregnant women. This result rejects the null hypothesis II(b), which is formulated for this purpose. The reason could be that the graduate pregnant women know the basic concepts of pregnancy and delivery much better than the uneducated pregnant women who are totally dependent on the elders and friends for the information.

The above table indicates that there is no significant difference in the level of stress between uneducated and postgraduate pregnant women. This result accepts the null; hypothesis 5(c), which is formulated for this purpose. The reason could that during pregnancy, the physical problems and risk factors are same for highly educated and uneducated pregnant women. In uneducated women, the more stress is may be due to lack of proper knowledge about pregnancy and delivery. While in highly educated women, the stress is more because of lack of practical knowledge though they have good bookish knowledge.

The above table indicates that there is significant difference in the level of stress between school level and graduate level of pregnant women. This result rejects the null hypothesis 5(d), which is formulated for this purpose. The reason could be that the school level educated pregnant women are afraid of many syndromes during pregnancy due to lack of knowledge and sometimes lack of confidence.

The above Table indicates that there is no significant difference in the level of stress between school level and postgraduate level of pregnant women. This result accepts the null hypothesis 5(e), which is formulated for this purpose. In this finding though the reason for the stress must be different but both are highly stressed. School level educated pregnant women have no or very less knowledge about the subject of pregnancy. Highly educated women seek the scientific reasons behind every physical and emotional change during pregnancy. Sometimes it confuses them more and increases their stress.

The Table 4.6 indicates that there is no significant difference in the level of stress between graduate level and postgraduate level of pregnant women. This result accepts the null hypothesis 5(f), which is formulated for this purpose. The reason may be that their knowledge is only through the books on the subject of pregnancy and delivery. Sometimes knowing only the crude factors without scientific reasons and how to handle them confuses them more.

6. Type of Family

The relationship between type of family and stress are observed in Table 4.7.

Table—4.7

The Stress Level Among the Pregnant Women Belonging to Joint Family and Nuclear Family

Sl. No.	Group	Mean	SD	t-value
1.	Joint family	397.79	40.52	0.09+
2.	Nuclear family	398.95	45.48	

\+ Not significant at 0.01 level.

The Table 4.7 indicates that there is no significant difference in the level of stress between pregnant women belonging to joint and nuclear family.

The reason could be since the phenomenon of stress is a common element among the pregnant women, because of the pregnancy and delivery involves different risk elements for the life of the pregnant women. So it is natural expectation that pregnant women have stress irrespective of type of the family.

7. Socio-economic Status

The relationship between socio-economic status and stress is shown in Table 4.8.

Table—4.8

The Stress Level Among the Pregnant Women Belonging to Different Types of Socio-economic Status

Sl. No.	Group	Mean	SD	F-value
1.	Lower	404.44	25.63	
2.	Middle	395.21	55.41	0.61+
3.	Higher	388.18	52.30	

\+ Not significant at 0.01 level.

The Table 4.8 indicates that there is no significant difference in the level of stress among pregnant women irrespective of their socio-economic status. Socio-economic status plays very minor role in stressed pregnant women. Because of the main conditions, which really affect the pregnant women, i.e., physical problems during pregnancy fear of labour pain, anxieties, for coping with the newborn almost similar.

5

MAJOR FINDINGS AND CONCLUSIONS

The present study was examined the level of prenatal stress among pregnant women. Earlier studies, though few in number have established that pregnant women with stress would undergo certain changes in their attitudes and behaviour. The study was conducted at the Southern Railway Hospital, Perambur, Chennai. This Centre renders medical and health care services to out and in patients of different types health problems. In this hospital, the clinical laboratory for diagnosing pregnancy conformation, counselling and HIV testing facilities are provided on weekdays only. About 5 to 10 pregnant women visit the hospital daily.

The study was conducted on a sample of 50 pregnant women at the Southern Railway Hospital, Perambur, Chennai. The sample included those who had undergone pregnancy testing and were confirmed and those who referred to integrated counselling and HIV testing. The investigator approached these pregnant women and requested them to participate in this study. The patients were made to feel at ease and an initial rapport was established. The questionnaire was given to pregnant women and they were asked to fill them as per the instructions.

MAJOR FINDINGS

The major findings of the study are the following ones:

1. There was no difference significant in the level of stress among pregnant women irrespective of their age.
2. Significant differences were observed in the level of stress among pregnant women in regard to number of pregnancies.
3. The level of stress among pregnant women had not significant influence on their months of pregnancy.
4. There were no significant differences in the level of stress among working and non-working pregnant women.
5. Significant differences were observed in the level stress among pregnant women irrespective of their level education.
6. No significant differences were noted in the level of stress among pregnant women belonging to joint and nuclear family.
7. There were no significant differences in the levels of stress among pregnant women irrespective of their socio-economic status.

DISCUSSION

The results of the study lead to the following conclusions:

1. In the age group of below 25 years, the reason may be their lack of mental maturity and fear of labour pain. In the second age group of 25 to 30 years of age, though they are physically and mentally mature the fears and in some cases lack of social support can increase their stress in pregnancy. In the above 30 years of age group, the chances of complicated pregnancy are more compare to younger age group.
2. The reason for the stress in the 1st pregnancy may be due to the anxiety for the condition 'pregnancy' and lack of knowledge and experience about it. Same way, in 2nd pregnancy, though they know the process yet the syndrome occurring during the pregnancy period and the outcome of labour make them develop stress. 3rd time pregnant women are more stressed.

The reason could be that nowadays the third baby increases the pressure on the women's health as well as on the family's finances.

3. It may be the same fears, anxieties and risk involved in pregnancy. In all stages of pregnancy, common syndromes can occur, i.e., pain in lower back, cramps, oedema, insomnia and nightmares, varicose veins etc.

4. Working pregnant women have to cope up with the work pressure as well as the physical change during the pregnancy. Non-working pregnant women do not have that extra income and syndromes occurring during the pregnancy are same for them. So the stress level is same in pregnant women irrespective of their job status.

5. The educated pregnant women know the basic concepts of pregnancy and delivery much better than the uneducated pregnant women who are totally dependent on the elders and friends for the information. During pregnancy, the physical problems and risk factors are same for highly educated and uneducated pregnant women. In uneducated women, the more stress is may be due to lack of proper knowledge about pregnancy and delivery. While in highly educated women, the stress is more because of lack of practical knowledge though they have good bookish knowledge.

6. The reason could be since the phenomenon of stress is a common element among the pregnant women, because of the pregnancy and delivery involves different risk elements for the life of the pregnant women. So it is natural expectation that pregnant women have stress irrespective of type of the family.

7. Socio-economic status plays very minor role in stressed pregnant women. Because the main conditions, which really affect the pregnant women, i.e., physical problems during pregnancy, fear of labour pain, anxieties, for coping with the newborn almost similar

BIBLIOGRAPHY

Allan Cameraon (2002), Professional Manual of Stress Assessment Questionnaire, *The Test Agency,* UK.

Arizmendi, T.G. (1987), Stressful Events Related to Pregnancy and Post Partum, *Journal Psychosom,* 31(6): 743-56.

Austin, M.P. (2003), Psychosocial Adjustment and Management of Depression and Anxiety in Pregnancy. *Australian Family Physician*, 32(3) 119-26.

Barker, D.P.J. (1995), Fetal Origins of Coronary Heart Disease, *British Medical Journal,* 311: 171-174.

Casear, P. (1993), Old and New Facts About Perinatal Brain Development, *Journal of Child Psychology and Psychiatry*, 1993, 314: 101-09.

Clements, A.D. (1992), The Incidence of Attention Deficit–Hyperactivity Disorder in Children whose Mothers Experienced Extreme Psychological Stress, *Georgia Educational Researcher,* 91: 1-14.

Cnattingius, S., Granath, F. Petersson G. and Harlow. L. (1999), The Influence of Gestational Age and Smoking Habits on the Risk of Subsequent Preterm Deliveries, *New England Journal of Medicine,* 341: 943-48.

Copper, R.L., Goldenberg, R.L., Das, A. Elder , N., Swain, M., Norman, G., Ramsey, R., Cotroneo, P., Collins , B.A., Johnson, F., Jones, P., and Meier, A.M. (1996), The Preterm Prediction Study: Maternal Stress is Associated with Spontaneous Preterm Birth at Less than Hirty-five Weeks' Gestation, *American Journal of Obstetrics and Gynecology*, 175: 1286-92.

Creasy, R.K. (1991), Lifestyle Influences on Prematurity, *Journal of Developmental Physiology*, 15: 15-20.

Day, N.L. and Richardson, G.A. (1991), Prenatal Alcohol Exposure: A Continuum of Effects, *Seminar of Perinatology*, 15: 271-79.

Dragonas, T. (1998) Early Childhood Education, *Clnical Psychology Review,* 189(2): 127-42.

Dunkel-Schetter, C. (1998), Maternal Stress and Preterm Delivery, *Prenatal and Neonatal Medicine,* 3: 39-42.

Eberhard-Gran, M. (2005), Treating Mood Disorders during Pregnancy, *Drug Safety,* 28(8): 695-706.

Eyler, F.D., Behenke, M., Conlon, M., Woods, N.S. and Wobie, K. (1998), Birth Outcome from a Prospective, Matched Study of Prenatal Crack/Cocaine Use: II. Interactive and Dose Effects on Neurobehavioural Assessment, *Pediatrics*, 1998, 101: 237-41.

Faden, V.B., Graubard, B.I. and Dufour, M. (1997). The Relationship of Drinking and Birth Outcome in a US National Sample of Expectant Mothers *Paediatric and Perinatal Epidemiology*, 11: 167-80.

Flynn, H.A. (2006) Rates and Predictors of Depression Treatment Among Pregnant Women in Hospital Affiliated. *General Hospital Psychiatry,* 28(4): 289-95.

Forde, R. (1992). Pregnant Women's Ailments and Psychosocal Conditions, *Family Practitioner,* 9(3): 270-73.

Geogas, J. and Giakoumaki, E. (1984), Psychological Stress and its Relation to Obstetrical Complications, Psychotherapy, *Psychosomatic,* 41(4): 200-06.

Haglund, L.J. (1998), The Perinatal Assessment of Psychosocial Risk, *Clinical Perinatal* 25(2): 417-52.

Hjaimarson, O., Hagberg, B. and Hagberg, G. (1998), Epidemiologic Panorama of Brain Impairments and Causative Factors: Swedish Experiences, In: Kubli, F., Patel, N., Schmidt, W. and Linderkamp, O. (Ed.), *Perinatal Events and Damage in Serving Children*, Berlin/Newyork: Springer-Verlag, pp. 28-38.

Hoffman, N.S. (1990), Stress Factors Related to Antenatal Testing During High Risk Pregnancy, *Journal Perinatal,* (10)2: 195-97.

Joffee, H. (1989), Emotional Factors in Pregnancy, *Australian Family Physician,* 18(5): 493, 496-97.

Kemp, V.H. (1986), Psycho-Social Impact of a High Risk Pregnancy on the Family, *Journal Obstetrics Gynecology Neonatal Nurse,* 15(3): 232-36.

Kirby, R.S. and Brewster, M.A. (1995), Early Childhood Surveillance of Developmental Disorders by a Birth Defects Surveillance System: Methods, Prevalence Comparisons, and Mortality Patterns, *Journal of Developmental Behavioural Pediatrics,* 16: 318-26.

Klock, S.C. (2000), Psychological Status of Invitro Fertilization Patients During Pregnancy, *Fertil Steril,* 73(6): 1159-64.

Lieberman, E., Torday, J., Barbieri, R., Cohen, A., Van Vunacis, H. and Weiss, S.T. (1992), Association of Intrauterine Cigarette Smoke Exposure with Indices of Fetal Lung Maturation, *Obstretrics and Gynecology,* 79: 564-70.

Maritz, G.S., Woolward, K.M and du Toit, G. (1993), Maternal Nicotine Exposure During Pregnancy and Development of Emphysema – Like Damage in the Offspring, *South African Medical Journal*, 83, 195-98.

Muhajarine, N. (1999), Physical Abuse During Pregnancy Prevalence and Risk Factors, *CMAJ,* 160(7): 1007-11.

Oakley, A. (1992), Measuring the Effectiveness of Psycho-social Interventions in Pregnancy, *International Journal Technology Assess Health Care,* pp. 129-38.

O'brien, L. (2007), Study of Depression, Anxiety, Irritability and Stress in Pregnancy, *Journal Psychiatry Practitioner,* 13(1): 33-39.

Pond, E.F. (1992), A Comparison Between Adolescent and Adult Women on Prenatal Anxiety and Self Confidence, *Journal Maternity Child Nurse*, 20(1): 11-20.

O'Callaghan, M.J., Williams, G.M., Andersen, M.J., Bor, W. and Najman, J.M. (1997), Obstetric and Perinatal Factors a Predictors of Child Behaviour at 5 Years, *Journal of Paediatrics and Child Health*, 33: 497-503.

Ravelli, A.C.J., Van der Meulen, J.K.P., Michels, R.P.J., Osmond, C., Barker, D.J.P., Hales, C.N. and Bleker, O.P. (1998), Glucose Tolerance in Adults After Prenatal Exposure to Famine, *The Lancet*, 351:173-77.

Robers, R.E., Attkisson, C.C. and Rosenblatt A. (1998), Prevalence of Psychopathology Among Children and Adolescents, *American Journal of Psychiatry*, 155: 715-25.

Sammons, L.N. (1990), Psychological Aspects of Second Pregnancy, *NAACOGS Clinical Issue Perinatal Women's Health Nurse*,1(3): 317-24.

Seifert, K.L. and Hoffnung, R. J. (1997), *Child and Adolescent Development*, Boston: Houghton Mifflin Company.

Sokol, R.J. and Clarren, S.K. (1989), *Guidelines for Use of Terminology Describing the Impact of Prenatal Alcohol in Offspring*, Alcohol and Clinical Experimental Research.

Tam, LW. (1998), Psychological Aspects of Pregnancy in the Military, *Military Medical,* 163(6): 408-12.

Taylor, A., Fisk, N.M. and Glover, V. (2000), Mode of Delivery and Subsequent Stress Response, *The Lancet*, 3555: 120.

Tuthill, D.P., Stewart, J.H., Coles, E.C., Andrews, J. and Cartlidge, P.H. (1999), Maternal Cigarette Smoking and Pregnancy Outcome, *Pediatrics and Prenatal Epidemiology*, 13: 245-53.

Uljas, H., Rautava, P., Helenius, H. and Sillanpaa, M. (1999), Behaviour of Finnish 3-year old Children: 1: Effects of Sociodemographic Factors, Mother's Health, and Pregnancy Outcome: *Developmental Medicine and Child neurology*: 41: 412-19.

Visser, G.H.A. and Narayan, H. (1996), The Problem of Increasing Severe Neurological Morbidity in Newborn Infants: Where Should the Focus be, *Prenatal Neurological Medicine*, 1: 12-15.

Wadhawa, P.D., Sandman, C.A, Porto, M., Dunkel-Schetter, C., and Garite, T.J. (1993), The Association Between Prenatal Stress and Infant Birth Weight and Gestational Age at Birth: A Prospective Investigation, *American Journal of Obstetrics and Gynecology*, 169: 858-65.

Weinstocks, M. (1997), Does Prenatal Stress Impair Coping and Regulation of Hypothalamic – Pituitary- Adrenal Axis, *Neuroscience and Biobehavioural Reviews*, 21: 1-10.

Williamson, P. (1981), Stress and Coping in First Pregnancy, *Journal of Family Practitioner,* 13(5): 629-35.

REFERENCES

Bhaskara Rao, Digumarti (1994), *Scientific Aptitude,* New Delhi: Ashish Publishing House, ISBN 81-7024-658-X.

Bhaskara Rao, Digumarti (1995), *Animal Kingdom,* New Delhi: Discovery Publishing House, ISBN 81-7141-274-2.

Bhaskara Rao, Digumarti (1995), *Batracology,* New Delhi: Discovery Publishing House, ISBN 81-7141-279-3.

Bhaskara Rao, Digumarti (1997), *Scientific Attitude,* New Delhi: Discovery Publishing House, ISBN 81-7141-381-1.

Bhaskara Rao, Digumarti (1996), *Scientific Attitude vis-à-vis Scientific Aptitude,* New Delhi: Discovery Publishing House, ISBN 81-7141-308-0.

Bhaskara Rao, Digumarti (2004), *Scientific Attitude, Scientific Aptitude and Achievement,* New Delhi: Discovery Publishing House, ISBN 81-7141-781-7.

Bhaskara Rao, Digumarti (2004), *Educational Administration,* New Delhi: Discovery Publishing House, ISBN 81-7141-842-2.

Bhaskara Rao, Digumarti (2004), *Issues in School Education,* New Delhi: Discovery Publishing House, ISBN 81-8356-025-3.

Bhaskara Rao, Digumarti, Editor (1996), *Encyclopaedia of Education For All,* 5 Volumes. New Delhi: APH Publishing Corporation, ISBN 81-7024-759-4 (set).

Vol. I *Education For All: The World Conference,* ISBN 81-7024-760-8

Vol. II *Education For All: The EPA-9 Summit,* ISBN 81-7024-761-6

Vol. III *Education For All: Quality Education For All,* ISBN 81-7024-762-6.

Vol. IV *Education For All: Planning and Monitoring,* ISBN 81-7024-763-4.

Vol. V *Education For All: The Indian Scenario,* ISBN 81-7024-764-0.

Bhaskara Rao, Digumarti, Editor (1996), *National Policy on Education,* 2 Volumes, New Delhi: Anmol Publications Pvt. Ltd, ISBN 81-7488-323-1.

Bhaskara Rao, Digumarti, Editor (1996), *Global Perceptions on Peace Education,* 3 Volumes, New Delhi: Discovery Publishing House, ISBN 81-7141-319-6.

Bhaskara Rao, Digumarti, Editor (1997), *Education for the 21st Century,* New Delhi: Discovery Publishing House, ISBN 81-7141-389-7.

Bhaskara Rao, Digumarti, Editor (1997), *Reflections on Scientific Attitude,* New Delhi: Discovery Publishing House, ISBN 81-7141-319-6.

Bhaskara Rao, Digumarti, Editor (1997), *Success Story of a Primary Education Project,* New Delhi: APH Publishing Corporation, ISBN 81-7024-850-7.

Bhaskara Rao, Digumarti, Editor (1997), *World Food Summit,* New Delhi: Discovery Publishing House, ISBN 81-7141-386-2.

Bhaskara Rao, Digumarti, Editor (1997), *Care the Child,* 2 Volumes, New Delhi: Discovery Publishing House, ISBN 81-7141-394-3.

Bhaskara Rao, Digumarti, Editor (1998), *Earth Summit,* 2 Volumes, New Delhi: Discovery Publishing House, ISBN 81-7141-435-4.

Bhaskara Rao, Digumarti, Editor (1998), *Adolescence Education,* New Delhi: Discovery Publishing House, ISBN 81-7141-432-X.

Bhaskara Rao, Digumarti, Editor (1998), *Community and School Nutrition Education,* New Delhi: Discovery Publishing House, ISBN 81-7141-435-4.

Bhaskara Rao, Digumarti, Editor (1998), *District Primary Education Programme,* New Delhi: Discovery Publishing House, ISBN 81-7141-396-X.

Bhaskara Rao, Digumarti, Editor (1998), *National Policy on Education: Towards an Enlightened and Humane Society,* New Delhi: Discovery Publishing House, ISBN 81-7141-426-5.

Bhaskara Rao, Digumarti, Editor (1998), *Reforming School Education,* New Delhi: Discovery Publishing House, ISBN 81-7141-403-6.

Bhaskara Rao, Digumarti, Editor (1998), *Teacher Education in India,* New Delhi: Discovery Publishing House, ISBN 81-7141-406-0.

Bhaskara Rao, Digumarti, Editor (1998), *World Summit for Social Development,* New Delhi: Discovery Publishing House, ISBN 81-7141-420-6.

Bhaskara Rao, Digumarti, Editor (1999), *International Encyclopaedia of AIDS,* 11 Volumes, New Delhi: Discovery Publishing House, ISBN 81-7141-522-6 (set).

Vol. 1 *Introduction to HIV/AIDS*, ISBN 81-7141-523-7.

Vol. 2 *HIV/AIDS – Issues and Challenges*, 2 Parts, ISBN 81-7141-524-5.

Vol. 3 *HIV/AIDS – Socio Economic Realities*, ISBN 81-7141-524-3.

Vol. 4 *HIV/AIDS – Law Ethics and Human Rights*, 2 Parts, ISBN 81-7141-526-1.

Vol. 5 *AIDS and NGOs*, ISBN 81-7141-527-X.

Vol. 6 *AIDS and Home Care*, ISBN 81-7141-528-8.

Vol. 7 *STD Case Management*, ISBN 81-7141-529-6.

Vol. 8 *HIV/AIDS Prevention and Care – Teaching Modules for Nurses and Midwives*, ISBN 81-7141-530-X.

Vol. 9 *HIV Prevention Education for Educational Institutions*, ISBN 81-7141-531-8.

Vol.10 *Instructional Modules for AIDS Education*, ISBN 81-7141-532-6.

Vol.11 *School Health Education to Prevent AIDS and STD – A Package for Curriculum Planners*, ISBN 81-7141-533-4.

Bhaskara Rao, Digumarti, Editor (2000), *International Encyclopaedia of Human Rights*, 7 Volumes in 13 Parts, New Delhi: Discovery Publishing House, ISBN 81-7141-567-9 (set).

Vol. 1 *International Instruments of Human Rights*, 2 Parts, ISBN 81-7141-569-4.

Vol. 2 *Regional Instruments of Human Rights*, ISBN 81-7141-604-7.

Vol. 3 *Human Rights and the United Nations*, 2 Parts. ISBN 81-7141-605-5.

Vol. 4 *Fact Files of Human Rights*, 3 Parts. ISBN 81-7141-606-3.

Vol. 5 *Study Stories of Human Rights*, 3 Parts. ISBN 81-7141-607-3.

Vol. 6 *International Meetings on Human Rights*, 2 Parts. ISBN 81-714-608-X.

Vol. 7 *Professional Training in Human Rights*, ISBN 81-7141-609-8.

Bhaskara Rao, Digumarti, Editor (2000), *International Encyclopaedia of Science and Technology Education*, 11 Volumes, New Delhi: Discovery Publishing House, ISBN 81-7141-548-2 (set).

Vol. 1 *Science and Technology Education*, ISBN 81-7141-568-7.

Vol. 2 *Science Education in Developing Countries*, ISBN 81-7141-569-9.

Vol. 3 *Organizational Structure of Science*, ISBN 81-7141-570-9.

Vol. 4 *Science Education in Asia and the Pacific*, ISBN 81-7141-571-7

Vol. 5 *Science and Technology Education For All*, ISBN 81-7141-572-5.

Vol. 6 *Values, Ethics, Talent and Girls in Science and Technology Education*, ISBN 81-7141-573-3.

Vol. 7 *Popularization of Science and Technology Education*, ISBN 81-7141-574-1.

Vol. 8 *Science, Power and Society*, ISBN 81-7141-575-X.

Vol. 9 *Information Technology*, ISBN 81-7141-576-8.

Vol.10 *Teacher Training in Science and Technology Education*, ISBN 81-7142-577-6.

Vol. 11 *Teacher Training in Science and Technology: A Curriculum Framework*. ISBN 81-7141-578-4.

Bhaskara Rao, Digumarti, Editor (2000). *Education For All: Achieving the Goal*, 3 Volumes. New Delhi: APH Publishing Corporation. ISBN 81-7648-152-1 (set).

Vol. I *The Global Consensus*, ISBN 81-7648-155-6.

Vol. II *Mid-Decade Review Reports of Regional Seminars,* ISBN 81-7648- 154-8.

Vol. III *Issues and Trends,* ISBN 81-7648-155-6.

Bhaskara Rao, Digumarti, Editor (2001), *Nuclear Materials: Issues and Concerns*, 2 Volumes, New Delhi: Discovery Publishing House, ISBN 81-7141-611-X.

Bhaskara Rao, Digumarti, Editor (2001), *Distance Education in Different Countries,* New Delhi: APH Publishing Corporation, ISBN 81-7648-229-3.

Bhaskara Rao, Digumarti, Editor (2001), *Decentralised Management of Education: Management of Education in Panchayati Raj and Municipal Bodies,* New Delhi: Discovery Publishing House, ISBN 81-7141-617-9.

Bhaskara Rao, Digumarti, Editor (2001), *Electrochemistry for Environmental Protection,* New Delhi: Discovery Publishing House, ISBN 81-7141-619-5.

Bhaskara Rao, Digumarti, Editor (2001), *Global Educational Studies,* New Delhi: Discovery Publishing House, ISBN 81-7141-616-0.

Bhaskara Rao, Digumarti, Editor (2001), *Global Synthesis of Educational Assessment,* New Delhi: Discovery Publishing House, ISBN 81-7141-613-6.

Bhaskara Rao, Digumarti, Editor (2001), *Jomtein Decade of Education,* New Delhi: Discovery Publishing House, ISBN 81-7141-618-7.

Bhaskara Rao, Digumarti, Editor (2001), *World Conference on Education for All,* New Delhi: APH Publishing Corporation, ISBN 81-7141-274-9.

Bhaskara Rao, Digumarti, Editor (2001), *World Conference on Higher Education,* New Delhi: Discovery Publishing House, ISBN 81-7141-610-1.

Bhaskara Rao, Digumarti, Editor (2001), *World Conference on Science,* New Delhi: Discovery Publishing House, ISBN 81-7141-612-8.

Bhaskara Rao, Digumarti, Editor (2003), *Inspiring Experiences in Teacher Education,* New Delhi: Discovery Publishing House, ISBN 81-7141-656-X.

Bhaskara Rao, Digumarti, Editor (2003), *International Studies in Education*, 3 Volumes, New Delhi: Discovery Publishing House, ISBN 81-7141-647-0.

Bhaskara Rao, Digumarti, Editor (2003), *Military Conversion: Impact on Science and Technology,* New Delhi: Discovery Publishing House, ISBN 81-7141-578-4.

Bhaskara Rao, Digumarti, Editor (2003), *United Nations Millennium Summit,* New Delhi: Discovery Publishing House, ISBN 81-7141-632-2.

Bhaskara Rao, Digumarti, Editor (2003), *World Assembly on Aging,* New Delhi: Discovery Publishing House, ISBN 81-7141-637-3.

Bhaskara Rao, Digumarti, Editor (2003), *World Conference on Human Rights,* New Delhi: Discovery Publishing House, ISBN 81-7141-661-6.

Bhaskara Rao, Digumarti, Editor (2003), *World Education Forum,* New Delhi: Discovery Publishing House, ISBN 81-7141-639-X.

Bhaskara Rao, Digumarti, Editor (2003), *Education, Employment and Human Resource Development*, New Delhi: Discovery Publishing House, ISBN 81-7141- 681-0.

Bhaskara Rao, Digumarti, Editor (2003), *Successful Schooling*, New Delhi: Discovery Publishing House, ISBN 81-7141-677-2.

Bhaskara Rao, Digumarti, Editor (2003), *European Education and Teachers*, New Delhi: Discovery Publishing House, ISBN 81-7141-702-7.

Bhaskara Rao, Digumarti, Editor (2003), *Teachers in a Changing World,* New Delhi: Discovery Publishing House, ISBN 81-7141-694-2.

Bhaskara Rao, Digumarti, Editor (2004), *International Guidelines on Open and Distance Teacher Education*, New Delhi: Discovery Publishing House, ISBN 81-7141-777-9.

Bhaskara Rao, Digumarti, Editor (2004), *Adult Learning in the 21st Century*, New Delhi: Discovery Publishing House, ISBN 81-7141-797-3.

Bhaskara Rao, Digumarti, Editor (2004), *Educational Practices: Research and Recommendations*, New Delhi: Discovery Publishing House, ISBN 81-7141-835-X.

Bhaskara Rao, Digumarti, Editor (2004), *General Secondary Education In the 21st Century*, New Delhi: Discovery Publishing House. ISBN 81-7141-885-6.

Bhaskara Rao, Digumarti, Editor (2004), *International Encyclopaedia of Learning to Live Together*, 4 Volumes, New Delhi: Discovery Publishing House, ISBN 81-7141-848-1.

Vol. 1 *International Conference on Learning to Live Together.*

Vol. 2 *Globalization and Living Together.*

Vol. 3 *Curriculum for Learning to Live Together.*

Vol. 4 *Science Education for the Contemporary Society.*

Bhaskara Rao, Digumarti, Editor (2004), *Reforming Secondary Education,* New Delhi: Discovery Publishing House, ISBN 81-7141-843-0.

Bhaskara Rao, Digumarti, Editor (2004), *Human Rights Education,* New Delhi: Discovery Publishing House, ISBN 81-7141-882-1.

Bhaskara Rao, Digumarti, Editor (2004), *United Nations Decade for Human Rights Education*, New Delhi: Discovery Publishing House, ISBN 81-7141-887-2.

Bhaskara Rao, Digumarti, Editor (2004), *Technical and Vocational Education and Training in the 21st Century,* New Delhi: Discovery Publishing House, ISBN 81-7141-984-4.

Bhaskara Rao, Digumarti, Editor (2005), *Encyclopaedia of Education For All,* 5 Volumes. New Delhi: Discovery Publishing House.

Bhaskara Rao, Digumarti and B.S.V. Dutt, Editors (2003), *Education: Programmes and Policies,* New Delhi: APH Publishing Corporation. ISBN 81-7648-470-9.

Bhaskara Rao, Digumarti, C.A.P. Swamy and B.S.V. Dutt (1997), *Self-Evaluation in Student Teaching,* New Delhi: Discovery Publishing House, ISBN 81-7141-374-9.

Bhaskara Rao, Digumarti and D. Naresh Kumar (2004), *School Teacher Effectiveness,* New Delhi: Discovery Publishing House, ISBN 81-7141-782-5.

Bhaskara Rao, Digumarti and D. Sridhar (2002), *Job Satisfaction of School Teachers*, New Delhi: Discovery Publishing House, ISBN 81-7141-652-7.

Bhaskara Rao, Digumarti, C. Sridevi and K. Vijaya (1995), *Achievement in Social Studies,* New Delhi: Discovery Publishing House, ISBN 81-7141-281-5.

Bhaskara Rao, Digumarti and Digumarti Pushpa Latha (1994), *Achievement in Biology,* New Delhi: Discovery Publishing House, ISBN 81-7141-264-5.

Bhaskara Rao, Digumarti and Digumarti Pushpa Latha (1995), *Achievement in English,* New Delhi: Discovery Publishing House, ISBN 81-7141-283-1.

Bhaskara Rao, Digumarti and Digumarti Pushpa Latha (1994), *Achievement in Science,* New Delhi: Discovery Publishing House, ISBN 81-7141-280-7.

Bhaskara Rao, Digumarti and Digumarti Pushpa Latha (1995), *Achievement in Mathematics,* New Delhi: Discovery Publishing House, ISBN 81-7141-278-5.

Bhaskara Rao, Digumarti and Digumarti Pushpa Latha (2004), *Education for Women,* New Delhi: Discovery Publishing House, ISBN 81-7141-873-2.

Bhaskara Rao, Digumarti, Digumarti Pushpa Latha and Digumarthi Harshitha, Editors (2001), *Biological Warfare,* New Delhi: Discovery Publishing House, ISBN 81-7141-597-0.

Bhaskara Rao, Digumarti, Digumarti Pushpa Latha and Digumarthi Harshitha, Editors (2001), *Women as Educators,* New Delhi: Discovery Publishing House, ISBN 81-7141-602-0.

Bhaskara Rao, Digumarti and Digumarthi Harshitha (2004), *Adjustment of Adolescents,* New Delhi: APH Publishing House, ISBN 81-7648-836-8.

Bhaskara Rao, Digumarti and Digumarthi Harshitha, Editors (2001), *Education in India,* New Delhi: APH Publishing House, ISBN 81-7648-207-2.

Bhaskara Rao, Digumarti and Digumarti Pushpa Latha, Editors (1998), *International Encyclopaedia of Women*, 5 Volumes. New Delhi: Discovery Publishing House, ISBN 81-7141-410-9 (set).

Vol. 1 *Status of World's Women,* ISBN 81-7141-494-X.

Vol. 2 *Women, Education and Empowerment,* ISBN 81-7141-498-1.

Vol. 3 *Women Challenges and Advancement,* ISBN 81-7141-497-4.

Vol. 4 *Women and Family Health,* ISBN 81-7141-497-4.

Vol. 5 *Women and International Action,* ISBN 81-7141-498-2.

Bhaskara Rao, Digumarti, Digumarti Pushpa Latha and Digumarthi Harshitha, Editors (2001), *Assessing Learning Achievement,* New Delhi: Discovery Publishing House, ISBN 81-7141-601-2.

Bhaskara Rao, Digumarti, Digumarti Pushpa Latha and Digumarthi Harshitha, Editors (2001), *Energy Security,* New Delhi: Discovery Publishing House, ISBN 81-7141-598-9.

Bhaskara Rao, Digumarti, Digumarthi Harshitha and K.R.S. Sambasiva Rao, Editors (1999), *Advanced Biotechnology,* New Delhi: Discovery Publishing House, ISBN 81-7141-516-4.

Bhaskara Rao, Digumarti and K.R.S. Sambasiva Rao, Editors (1996), *Current Trends in Indian Education,* New Delhi: Discovery Publishing House, ISBN 81-7141-311-0.

Bhaskara Rao, Digumarti and D. Naresh Kumar (2004), *School Teacher Effectiveness*, New Delhi: Discovery Publishing House. ISBN 81-7141-782-5.

Bhaskara Rao, Digumarti and E. Sreekanth Babu (2004), *Educational Interests of School Students*, New Delhi: Discovery Publishing House, ISBN 81-7141-837-6.

Bhaskara Rao, Digumarti and K. Vijaya (1995), *A Text Book Evaluation,* Ambala Cantt: The Associated Publishers.

Bhaskara Rao, Digumarti and M.A. Fayaz (2004), *Problems of Primary School Drop-outs,* New Delhi: Discovery Publishing House, ISBN 81-7141- 834-1.

Bhaskara Rao, Digumarti and N.V.M. Mohana Rao (2002), *Problems of Mentally Handicapped Children*, New Delhi: Discovery Publishing House, ISBN 81-7141- 645-4.

Bhaskara Rao, Digumarti and S. Chandra Mohan (2002), *Sports Management*, New Delhi: APH Publishing House, ISBN 81-7648-467-9.

Bhaskara Rao, Digumarti and S.A. Khader (2004), *Problems of Private School Teachers,* New Delhi: Discovery Publishing Corporation, ISBN 81-7141-838-4.

Bhaskara Rao, Digumarti and S.A. Khader (2004), *School Education in India,* New Delhi: Discovery Publishing Corporation, ISBN 81-7141-849-X.

Bhaskara Rao, Digumarti and Sk. Johni Basha (2004), *Teachers' Population Education Awareness,* New Delhi: Discovery Publishing House, ISBN 81-7141-832-5.

Bhaskara Rao, Digumarti, V.V. Rao, V.V. Lakshmi and V.V. Krishna, Editors (1999), *Status and Advancement of Women,* New Delhi: APH Publishing Corporation, ISBN 81-7648-169-6.

Appala Naidu, P.Ch., Author and Digumarti Bhaskara Rao, editor (2007), *Feedback Methods and Student Performance,* New Delhi: Discovery Publishing House, ISBN 81-8356-284-1.

Babu, P.C., Author and Digumarti Bhaskara Rao, Editor (2004), *Flowers of Wisdom,* New Delhi: Discovery Publishing House, ISBN 81-7141-695-0.

Bujji Babu, K., Author and Digumarti Bhaskara Rao, Editor (2007), *Teaching Aptitude of Primary School Teachers,* New Delhi: Sonali Publications, ISBN 81-8411-083-9.

Amala, P. A. and Anupama, P., Authors and Digumarti Bhaskara Rao, Editor (2004), *History of Education,* New Delhi: Discovery Publishing House, ISBN 81-7141-860-0.

Bhagya Lakshmi, L., Author and Digumarti Bhaskara Rao, Editor (2000), *Reading and Comprehension,* New Delhi: Discovery Publishing House, ISBN 81-7141-543-1.

Bhasha, S.A., Author and Digumarti Bhaskara Rao, Editor (2004), *Methods of Teaching Geography,* New Delhi: Discovery Publishing House, ISBN 81-7141-807-4.

Bhuvaneswara Lakshmi, Gadde, Author and Digumarti Bhaskara Rao, Editor (2000), *Attitude Towards Science,* New Delhi: Discovery Publishing House, ISBN 81-7141-541-46.

Bhuvaneswara Lakshmi, G., Author and Digumarti Bhaskara Rao, Editor (2004), *Methods of Teaching Life Science,* New Delhi: Discovery Publishing House, ISBN 81-7141-804-X.

Bhuvaneswara Lakshmi, G. and K. Subba Rao, Authors and Digumarti Bhaskara Rao, Editor (2004), *Methods of Teaching Biology,* New Delhi: Discovery Publishing House, ISBN 81-7141-914-3.

Chary, K.V.N.B., Author and Digumarti Bhaskara Rao, Editor (2006), *Techniques of Teaching Physics,* New Delhi: Sonali Publications. ISBN 81-8411-046-4.

Chowdary, S.B.J.R. and Naga Raju, Authors and Digumarti Bhaskara Rao, Editor (2004), *Mastery of Teaching Skills,* New Delhi: Discovery Publishing House, ISBN 81-7141-861-9.

Dayakara Reddy, V. and Digumarti Bhaskara Rao, Editors (2006), *Value-Oriented Education,* New Delhi: Discovery Publishing House, ISBN 81-8356-051-2.

Devraj, T.A.S., Author and Digumarti Bhaskara Rao, Editor (1997), *Trace Analysis of Uranium and Thorium,* New Delhi: Discovery Publishing House, ISBN 81-7141-375-7.

Durga Rani, K., Author and Digumarti Bhaskara Rao, Editor (2000), *Educational Aspirations and Scientific Attitudes,* New Delhi: Discovery Publishing House, ISBN 81-7141-555-5.

Dutt, B.S.V. and Digumarti Bhaskara Rao (2001), *Empowering Primary Teachers,* New Delhi: Discovery Publishing House, ISBN 81-7141-615-2.

Dutt, B.S.V., Author and Digumarti Bhaskara Rao, Editor (2004), *Comparative Education,* New Delhi: Discovery Publishing House, ISBN 81-7141-912-7.

Ediger, Marlow and Digumarti Bhaskara Rao (1996), *Science Curriculum,* New Delhi: Discovery Publishing House, ISBN 81-7141-321-8.

Ediger, Marlow and Digumarti Bhaskara Rao (2000), *Teaching Mathematics Successfully,* New Delhi: Discovery Publishing House, ISBN 81-7141-552-0.

Ediger, Marlow and Digumarti Bhaskara Rao (2001), *Teaching Science Successfully,* New Delhi: Discovery Publishing House, ISBN 81-7141-600-4.

Ediger, Marlow and Digumarti Bhaskara Rao (2001), *Teaching Social Studies Successfully,* New Delhi: Discovery Publishing House, ISBN 81-7141-596-2.

Ediger, Marlow and Digumarti Bhaskara Rao (2002), *Philosophy and Curriculum,* New Delhi: Discovery Publishing House, ISBN 81-7141-631-4.

Ediger, Marlow and Digumarti Bhaskara Rao (2002), *Improving School Administration,* New Delhi: Discovery Publishing House, ISBN 81-7141-633-0.

Ediger, Marlow and Digumarti Bhaskara Rao (2002), *Elementary Curriculum,* New Delhi: Discovery Publishing House, ISBN 81-7141-658-6.

Ediger, Marlow and Digumarti Bhaskara Rao (2003), *Language Arts Curriculum,* New Delhi: Discovery Publishing House. ISBN 81-7141-657-8.

Ediger, Marlow and Digumarti Bhaskara Rao (2003), *Psychology and Curriculum,* New Delhi: Discovery Publishing House, ISBN 81-7141-691-8.

Ediger, Marlow and Digumarti Bhaskara Rao (2003), *Teaching Language Arts Successfully,* New Delhi: Discovery Publishing House, ISBN 81-7141-678-0.

Ediger, Marlow and Digumarti Bhaskara Rao (2003), *School Curriculum and Administration,* New Delhi: Discovery Publishing House, ISBN 81-7141-709-4.

Ediger, Marlow and Digumarti Bhaskara Rao (2003), *Teaching Mathematics in Elementary Schools,* New Delhi: Discovery Publishing House, ISBN 81-7141-687-X.

Ediger, Marlow and Digumarti Bhaskara Rao (2003), *Teaching Science in Elementary Schools,* New Delhi: Discovery Publishing House, ISBN 81-7141-698-5.

Ediger, Marlow and Digumarti Bhaskara Rao (2003), *School Curriculum and Administration,* New Delhi: Discovery Publishing House, ISBN 81-7141-709-4.

Ediger, Marlow and Digumarti Bhaskara Rao (2003), *Elementary Curriculum Improvement,* New Delhi: Discovery Publishing House, ISBN 81-7141-740-X.

Ediger, Marlow and Digumarti Bhaskara Rao (2004), *School Organisation,* New Delhi: Discovery Publishing House. ISBN 81-7141-843-0.

Ediger, Marlow and Digumarti Bhaskara Rao (2004). *Relevancy in Elementary Curriculum,* New Delhi: Discovery Publishing House, ISBN 81-7141-845-9.

Ediger, Marlow and Digumarti Bhaskara Rao (2005), *Quality School Education,* New Delhi: Discovery Publishing House, ISBN 81-8356-022-9.

Ediger, Marlow and Digumarti Bhaskara Rao (2006), *Successful School Education,* New Delhi: Discovery Publishing House, ISBN 81-8356-054-7.

Ediger, Marlow and Digumarti Bhaskara Rao (2006), *Successful School Administration,* New Delhi: Discovery Publishing House, ISBN 81-8356-046-6.

Ediger, Marlow and Digumarti Bhaskara Rao (2006), *Issues in School Curriculum,* New Delhi: Discovery Publishing House, ISBN 81-8356-052-0.

Ediger, Marlow and Digumarti Bhaskara Rao (2006), *Community College – Curriculum and Teaching,* New Delhi: Discovery Publishing House, ISBN 81-8356-053-9.

Ediger, Marlow and Digumarti Bhaskara Rao (2006), *Administration of Schools,* New Delhi: Discovery Publishing House, ISBN 81-8356-244-2.

Ediger, Marlow and Digumarti Bhaskara Rao (2006), *Reading Curriculum and Instruction,* New Delhi: Discovery Publishing House, ISBN 81-8356-266-3.

Ediger, Marlow and Digumarti Bhaskara Rao (2006), *Curriculum Organisation,* New Delhi: Discovery Publishing House, ISBN 81-8356-205-1.

Ediger, Marlow and Digumarti Bhaskara Rao (2006), *Curriculum of School Subjects,* New Delhi: Discovery Publishing House, ISBN 81-8356-207-8.

Ediger, Marlow, B.S.V. Dutt and Digumarti Bhaskara Rao (2003), *Teaching English Successfully,* New Delhi: Discovery Publishing House, ISBN 81-7141-707-8.

Ediger, Marlow and Digumarti Bhaskara Rao (2007), *School Science Education,* New Delhi: Discovery Publishing House. ISBN 81-8356-352-X.

Ediger, Marlow and Digumarti Bhaskara Rao (2007), *Language Arts Education,* New Delhi: Discovery Publishing House, ISBN 81-8356-333-3.

Elizabeth, M.E.S., Author and Digumarti Bhaskara Rao, Editor (2004), *Methods of Teaching English,* New Delhi: Discovery Publishing House, ISBN 81-7141-809-0.

Elizabeth, M.E.S., Author and Digumarti Bhaskara Rao, Editor (2004), *Acquisition of English Vocabulary,* New Delhi: Discovery Publishing House, ISBN 81-8356-075-X.

Fatima, Sk. Author and Digumarti Bhaskara Rao, Editor (2007), *Reasoning Ability of School Students,* New Delhi: Discovery Publishing House, ISBN 81-8356-330-9.

Fatima, Sk. and Digumarti Bhaskara Rao (2008), *Reasoning Ability of Adolescent Students,* New Delhi: Sonali Publications.

Gopala Krishna, M., Author and Digumarti Bhaskara Rao, Editor (2007), *Techniques of Teaching Physical Education,* New Delhi: Sonali Publications, ISBN 81-8411-044-8.

Gopala Krishna, M., Author and Digumarti Bhaskara Rao, Editor (2007), *Techniques of Teaching Education,* New Delhi: Sonali Publications, ISBN 81-8411-062-6.

Harshitha, Digumarthi, Author and Digumarti Bhaskara Rao, Editor (2004), *Methods of Teaching Information Technology,* New Delhi: Discovery Publishing House. ISBN 81-7141-805-8.

Harshitha, Digumarthi, Author and Digumarti Bhaskara Rao, Editor (2007), *Techniques of Teaching Computer Science,* New Delhi: Sonali Publications, ISBN 81-8411-036-7.

Indira Devi, Author and J. Prasanth Kumar and Digumarti Bhaskara Rao, Editors (2004), *Values in Language Text Books,* New Delhi: Discovery Publishing House, ISBN 81-7141-833-7.

Jalaja Kumari, C., Author and Digumarti Bhaskara Rao, Editor (2004), *Methods of Teaching Educational Technology,* New Delhi: Discovery Publishing House. ISBN 81-7141-810-4.

Jalaja Kumari, C., Author and Digumarti Bhaskara Rao, Editor (2007), *Job Satisfaction of Teachers,* New Delhi: Discovery Publishing House, ISBN 81-8356-329-5.

Janardhan Reddy, B., Author and Digumarti Bhaskara Rao, editor (2006), *Techniques of Teaching Sociology,* New Delhi: Sonali Publications, ISBN 81-8411-042-1.

Jayasree, K., Author and Digumarti Bhaskara Rao, Editor (1999), *Correlates of Socialisation,* New Delhi: Discovery Publishing House, ISBN 81-7141-517-2.

Jayasree, K., Author and Digumarti Bhaskara Rao, Editor (2004), *Methods of Teaching Science,* New Delhi: Discovery Publishing House, ISBN 81-7141-801-5.

John Babu, C., Author and T.J.R. Prasad, G.M. Madhukar and Digumarti Bhaskara Rao, Editors (2004), *Problem Solving in Mathematics,* New Delhi: APH Publishing Corporation, ISBN 81-7648-273-0.

Joseph Raju, B and G.A. Anitha, Authors and Digumarti Bhaskara Rao, Editor (2004), *Population Education,* New Delhi: Sonali Publications. ISBN 81-88836-31-3.

Lalitha, T., Author and K.S. Prabhakaram, D.S.N. Sastry and Digumarti Bhaskara Rao, Editors (2004), *Educational Philosophic Beliefs,* New Delhi: Discovery Publishing House, ISBN 81-7141-765-5.

Krishna, G., Author and Digumarti Bhaskara Rao, Editor (2006), *Techniques of Teaching Physical Education,* New Delhi: Sonali Publications, ISBN 81-8411-044-8.

Kumar Raja, G., Author and Digumarti Bhaskara Rao, Editor (2007), *Principles of Primary School,* New Delhi: Sonali Publications, ISBN 81-8411-054-5.

Lakshmi Kumari, V., Author and Digumarti Bhaskara Rao, Editor (2006), *Techniques of Teaching Home Science,* New Delhi: Sonali Publications, ISBN 81-8411-048-0.

Madhava, K., Author and Digumarti Bhaskara Rao, Editor (2008), *Personality of Adolescent Students,* New Delhi: Discovery Publishing House, ISBN 978-81-8356-262-1.

Madhu Bala, Jampala, Author and Digumarti Bhaskara Rao, Editor (2004), *Methods of Teaching Exceptional Children,* New Delhi: Discovery Publishing House, ISBN 81-7141-802-3.

Madhu Bala, Jampala, Author and Digumarti Bhaskara Rao, Editor (2007), *Adjustment Problems of Hearing Impaired,* New Delhi: Discovery Publishing House, ISBN 81-7141-831-7.

Marja, Talvi and Digumarti Bhaskara Rao, Editors (1996), *Educational Leadership and Social Changes,* New Delhi: Discovery Publishing House, ISBN 81-7141-320-X.

Naga Kumari, U., Author and Digumarti Bhaskara Rao, Editor (2008), *Science Process Skills of School Students,* New Delhi: Discovery Publishing House, ISBN 978-81-8356-263-8.

Nageswara Rao, S. and M. Srihari, Authors and Digumarti Bhaskara Rao, Editor (2004), *Guidance and Counselling,* New Delhi: Discovery Publishing House, ISBN 81-7141-840-6.

Nageswara Rao, S., Author and Digumarti Bhaskara Rao, Editor (2006), *Techniques of Teaching Psychology,* New Delhi: Sonali Publications, ISBN 81-8411-040-5.

Nageswara Rao, S. and P. Sridhar, Authors and Digumarti Bhaskara Rao, Editor (2004), *Methods and Techniques of Teaching,* New Delhi: Sonali Publications. ISBN 81-88836-33-8.

Nirmala Jyothi, M., Author and Digumarti Bhaskara Rao, Editor (2003), *Non-detention System in School Education,* New Delhi: Discovery Publishing House, ISBN 81-7141-654-3.

Padma Tulasi, G., Author and Digumarti Bhaskara Rao, Editor (2004), *Methods of Teaching Elementary Science,* New Delhi: Discovery Publishing House, ISBN 81-7141-871-6.

Pala Prasada Rao, V., Author and K.N. Rani and D. Bhaskara Rao, Editors (2004), *India Pakistan: Partition Perspectives in Indo English Novels,* New Delhi: Discovery Publishing House, ISBN 81-7141-871-6.

Pala Prasada Rao, V., Author and D. Bhaskara Rao, Editors (2008). *Functioning of Autonomous Colleges,* New Delhi: Discovery Publishing House, ISBN 978-81-8356-258-4.

Pitchi Reddy, M., Author and Digumarti Bhaskara Rao, Editor (2007), *Techniques of Teaching Social Sciences,* New Delhi: Sonali Publications, ISBN 81-8411-066-X.

Prasad Babu, B., Author and P. Madhu and Digumarti Bhaskara Rao, Editors (2006), *Psychological Adjustment and Well-being,* New Delhi: Discovery Publishing House, ISBN 81-8356-204-3.

Prasad Babu, B., Author and M.V.R. Raju and Digumarti Bhaskara Rao, Editors (2006), *Behavioural Problems of School Children,* New Delhi: Discovery Publishing House, ISBN 81-8356-206-X.

Prabhakaram, K.S., Author and Digumarti Bhaskara Rao, Editors (1998), *Concept Attainment Model in Mathematics Teaching,* New Delhi: Discovery Publishing House, ISBN 81-7141-424-9.

Prasanth Kumar, J., Author and Digumarti Bhaskara Rao, Editor (1998), *Effectiveness of Distance Education System,* New Delhi: Discovery Publishing House, ISBN 81-7141-437-0.

Prasanth Kumar, J., Author and Digumarti Bhaskara Rao, Editor (2004), *Methods of Teaching Civics,* New Delhi: Discovery Publishing House, ISBN 81-7141-806-6.

Prasanth Kumar, J., Author and G. Sundara Rao and Digumarti Bhaskara Rao, Editors (2000), *Open University Student Support Services,* New Delhi: Discovery Publishing House, ISBN 81-7141-550-4.

Raja Kumari, M.A. and D.R.S. Sundari, Authors and Digumarti Bhaskara Rao, Editor (2004), *Special Education,* New Delhi: Discovery Publishing House, ISBN 81-7141-846-5.

Raja Kumari, M.A. and D.R.S. Sundari, Authors and Digumarti Bhaskara Rao, Editor (2004), *Methods of Teaching Educational Psychology,* New Delhi: Discovery Publishing House. ISBN 81-7141-820-1.

Ramatulasamma, K., Author and Digumarti Bhaskara Rao, Editor (2002). *Job Satisfaction of Teacher Educators,* New Delhi: Discovery Publishing House, ISBN 81-7141-655-1.

Rama Krishnaiah, D., Author and Digumarti Bhaskara Rao, Editor (1998), *Job Satisfaction of College Teachers,* New Delhi: Discovery Publishing House, ISBN 81-7141-438-9.

Rama Kumar Ratnam, M.V., Author and Digumarti Bhaskara Rao, Editor (1998), *Dukkha: Suffering in Early Buddhism,* New Delhi: Discovery Publishing House, ISBN 81-7141-653-5.

Rama Krishna Prasad and P. Vide Sagar, Authors and Digumarti Bhaskara Rao, Editor (2004), *Methods of Teaching Physical Education,* New Delhi: Discovery Publishing House, ISBN 81-7141-868-6.

Rama Seshaiah, P. Author and Digumarti Bhaskara Rao, Editor (2004), *Methods of Teaching Home Science,* New Delhi: Discovery Publishing House, ISBN 81-7141-916-X.

Rama Swamy, K., Author and Digumarti Bhaskara Rao, Editor (2007), *Techniques of Teaching Environmental Science,* New Delhi: Sonali Publications. ISBN 81-8411-035-9.

Ramesh, A.R., Author and Digumarti Bhaskara Rao, Editor (2006), *Techniques of Teaching Commerce,* New Delhi: Sonali Publications, ISBN 81-8411-043-X.

Ramesh, Ghanta and Digumarti Bhaskara Rao, Editors (1998), *Environmental Education: Problems and Prospects,* New Delhi: Discovery Publishing House, ISBN 81-7141-423-0.

Ranga Rao, B., Author and Digumarti Bhaskara Rao, Editor (2007), *Techniques of Teaching Economics,* New Delhi: Sonali Publications, ISBN 81-8411-056-1.

Ranga Rao, R., Author and Digumarti Bhaskara Rao, Editor (2004), *Methods of Teacher Teaching,* New Delhi: Discovery Publishing House, ISBN 81-7141-812-0.

Rani, S.S., Author and Digumarti Bhaskara Rao, Editor (2006), *Techniques of Teaching Botany,* New Delhi: Discovery Publishing House, ISBN 81-8411-037-5.

Rathaiah, Lavu and Digumarti Bhaskara Rao, Editors (1996), *International Innovations in Education,* New Delhi: Discovery Publishing House, ISBN 81-7141-359-5.

Rathaiah, Lavu and Digumarti Bhaskara Rao (1997), *Achievement Correlates,* New Delhi: Discovery Publishing House, ISBN 81-7141-385-4.

Ravi Krishna, M., Author and Digumarti Bhaskara Rao, Editor (2004), *Examination System,* New Delhi: Discovery Publishing House, ISBN 81-7141-824-4.

Ravi Kumar, M., Author and Digumarti Bhaskara Rao, Editor (2004), *Methods of Teaching Computer Science,* New Delhi: Discovery Publishing House, ISBN 81-7141-823-6.

Rudramamba, B., Author and Digumarti Bhaskara Rao, Editor (2003), *Problems of Teaching,* New Delhi: APH Publishing Corporation, ISBN 81-7648-462-8.

Rudramamba, B. and V. Lakshmi Kumari, Authors and Digumarti Bhaskara Rao, Editor (2004), *Methods of Teaching Economics,* New Delhi: Discovery Publishing House. ISBN 81-7141-900-3.

Sambasiva Rao, P., Author and Digumarti Bhaskara Rao, Editor (2007), *Techniques of Teaching Psychology,* New Delhi: Sonali Publications, ISBN 81-8411-040-5.

Sanjeeva Rao, P.C., Author and Digumarti Bhaskara Rao, Editor (1996), *A Text Book of Geology,* New Delhi: Discovery Publishing House, ISBN 81-7141-313-7.

Santhanam, T., B. Prasad Babu and S. Sugandhi, Authors and Digumarti Bhaskara Rao, Editor (2007), *Children with Learning Disabilities,* New Delhi: Sonali Publications. ISBN 81-8411-077-4.

Santhanam, T., B. Prasad Babu and S. Sugandhi, Authors and Digumarti Bhaskara Rao, Editor (2008), *Learning Disabilities and Remedial Programmes,* New Delhi: Discovery Publishing House. ISBN 978-81-8356-257-7.

Sarala, M.M.O., Author and Digumarti Bhaskara Rao, Editor (2006), *Techniques of Teaching English,* New Delhi: Sonali Publications, ISBN 81-8411-047-2.

Satya Narayana, G., Author and Digumarti Bhaskara Rao, Editor (2008), *Attitude Towards Social Studies and Achievement in Social Studies,* New Delhi: Discovery Publishing House. ISBN 978-81-8356-261-4.

Satya Narayana, V., Author and Digumarti Bhaskara Rao, Editor (2001), *Physical Education, Social Attitudes and Leadership Qualities,* New Delhi: Discovery Publishing House, ISBN 81-7141-593-8.

Satya Narayana, P.V.V. and G. Krishna, Authors and Digumarti Bhaskara Rao, Editor (2004), *Curriculum Development and Management,* New Delhi: Discovery Publishing House, ISBN 81-7141-813-9.

Shamsuddin, Sk. and V. Dayakara Reddy, Authors and Digumarti Bhaskara Rao, Editor (2007), *Academic Achievement and Values,* New Delhi: Discovery Publishing House.

Singh, Y.C., author and Digumarti Bhaskara Rao, Editor (2006), *Techniques of Teaching Science,* New Delhi: Sonali Publications. ISBN 81-8411-041-3.

Sirisha Rani, S., Author and Digumarti Bhaskara Rao, Editor (2007), *Techniques of Teaching Botany,* New Delhi: Sonali Publications. ISBN 81-8411-037-5.

Sivaratnam Reddy, M., Author and Digumarti Bhaskara Rao, Editor (2004), *Creativity in College Students,* New Delhi: Discovery Publishing House, ISBN 81-7141-697-7.

Siva Lakshmi, G.V. and G.L. Subbaiah, Authors and Digumarti Bhaskara Rao, Editor (2004), *Methods of Teaching Environmental Science,* New Delhi: Discovery Publishing House, ISBN 81-7141-839-2.

Srinivas, G. and Digumarti Bhaskara Rao (2007), *Anxiety of Prospective Teachers,* New Delhi: Sonali Publications, ISBN 81-8411-084-7.

Srinivas, M. and I. Prasada Rao, Authors and Digumarti Bhaskara Rao, Editor (2004), *Methods of Teaching History*, New Delhi: Discovery Publishing House, ISBN 81-7141-803-1.

Srinivas Rao, P., Author and Digumarti Bhaskara Rao, Editor (2007), *Principles of Secondary School*, New Delhi: Sonali Publications, ISBN 81-8411-058-8.

Srinivasulu Reddy, M. and K.R.S. Sambasiva Rao, Authors and Digumarti Bhaskara Rao, Editor (1999), *A Text Book of Aquaculture*, New Delhi: Discovery Publishing House, ISBN 81-7141-482-6.

Srinivasa Rao, Mandalapu, Author and Digumarti Bhaskara Rao, Editor (2003), *Achievement Motivation and Achievement in Mathematics*, New Delhi: Discovery Publishing House, ISBN 81-7141-674-8.

Srihari, M., Author and Digumarti Bhaskara Rao, Editor (2003), *Values of Prospective Teachers*, New Delhi: Discovery Publishing House, ISBN 81-8356-328-7.

Subba Rao, K., Author and Digumarti Bhaskara Rao, Editor (2007), *School Education Policy*, New Delhi: Discovery Publishing House, ISBN 81-8356-285-X.

Subba Rao, K., Author and Digumarti Bhaskara Rao, Editor (2007), *Education Planning*, New Delhi: Sonali Publications, ISBN 81-8411-053-7.

Sudhakar Reddy, Y., Author and Digumarti Bhaskara Rao, Editor (2003), *Creativity in Adolescents*, New Delhi: Discovery Publishing House, ISBN 81-7141-659-4.

Sunil Kumar, K. and K. Rama Krishana, Authors and Digumarti Bhaskara Rao, Editor (2004), *Methods of Teaching Chemistry*, New Delhi: Discovery Publishing House, ISBN 81-7141-913-5.

Suneetha, G., Author and Digumarti Bhaskara Rao, Editor (2004), *Environmental Awareness of School Students*, New Delhi: Sonali Publications, ISBN 81-8411-085-5.

Sunita, E. and R. Sambasiva Rao, Authors and Digumarti Bhaskara Rao, Editor (2004), *Methods of Teaching Mathematics,* New Delhi: Discovery Publishing House. ISBN 81-7141-915-1.

Surya Madhava, I., Author and Digumarti Bhaskara Rao, Editor (2006), *Techniques of Teaching Geography,* New Delhi: Discovery Publishing House, ISBN 81-8411-034-0.

Surya Madhava, I., Author and Digumarti Bhaskara Rao, Editor (2007), *Techniques of Teaching Political Science,* New Delhi: Discovery Publishing House, ISBN 81-8411-061-8.

Swamy, K.R., Author and Digumarti Bhaskara Rao, Editor (2006), *Techniques of Teaching Environmental Science,* New Delhi: Discovery Publishing House, ISBN 81-8411-035-9.

Swarna Jyothi, K., Author and Digumarti Bhaskara Rao, Editor (2007), *Educational Research,* New Delhi: Sonali Publications. ISBN 81-8411-063-4.

Swarna Latha, C.D., and Digumarti Bhaskara Rao, Editors (2006), *Encyclopaedia of Biotechnology,* 5 Volumes, New Delhi: Discovery Publishing House, ISBN 81-8356-168-3.

Swarupa Rani, T. and J.R. Priyadarshini, Authors and Digumarti Bhaskara Rao, Editor (2004), *Educational Measurement and Evaluation,* New Delhi: Discovery Publishing House, ISBN 81-7141-859-7.

Vanaja, M., Author and Digumarti Bhaskara Rao, Editor (1999), *Inquiry Training Model,* New Delhi: Discovery Publishing Housc, ISBN 81-7141-515-6.

Vanaja, M., Author and Digumarti Bhaskara Rao, Editor (2004), *Methods of Teaching Physics,* New Delhi: Discovery Publishing House, ISBN 81-7141-867-8

Valeri V. Koustiouk, Author and Digumarti Bhaskara Rao, Editor (2002), *A Text Book of Cryogenics,* New Delhi: Discovery Publishing House, ISBN 81-7141-642-X.

Vamsi Krishna, V., Author and Digumarti Bhaskara Rao, Editor (2004), *School Psychology,* New Delhi: Discovery Publishing House, ISBN 81-7141-880-5.

Veena Kumari, Balusu and Digumarti Bhaskara Rao (1996), *Operation Black Board,* New Delhi: APH Publishing Corporation. ISBN 81-7024-711-X.

Veena Kumari, Balusu, Author and Digumarti Bhaskara Rao, Editor (2004), *Methods of Teaching Social Studies,* New Delhi: Discovery Publishing House, ISBN 81-7141-899-6.

Veena Kumari, Balusu, Author and Digumarti Bhaskara Rao, Editor (2000), *Psycho-Social Correlates of Achievement,* New Delhi: Discovery Publishing House, ISBN 81-7141-547-4.

Venkata Rao, B., Author and Digumarti Bhaskara Rao, Editor (2007), *Techniques of Teaching Chemistry,* New Delhi: Sonali Publications, ISBN 81-8411-057-X.

Venkata Rao, P. and Digumarti Bhaskara Rao (1989), *A Text Book of Zoology – Junior Intermediate,* Guntur: Vignan Publishers.

Venkata Rao, P. and Digumarti Bhaskara Rao (1989), *A Text Book of Zoology – Senior Intermediate,* Guntur: Vignan Publishers.

Venkateswara Rao, V., Author and Digumarti Bhaskara Rao, Editor (2004), *Problems of Education,* New Delhi: Discovery Publishing House, ISBN 81-7141-841-4.

Venkateswara Rao, V., V. Vijaya Lakshmi and V. Vamsi Krishna, Authors and Digumarti Bhaskara Rao, Editor (2004), *Education For All,* New Delhi: Sonali Publications. ISBN 81-88836-30-3.

Venkateswara Rao, V., V. Vijaya Lakshmi and V. Vamsi Krishna, Authors and Digumarti Bhaskara Rao, Editor (2004), *Education in India,* New Delhi: Sonali Publications. ISBN 81-88836-858-9.

Venkateswara Reddy, L. and Narayana, M. L., Authors and Digumarti Bhaskara Rao, Editor (2004). *Education for Dalits,* New Delhi: Discovery Publishing House. ISBN 81-7141-872-4.

Venkateswara Reddy, L. and Narayana, M. L, Authors and Digumarti Bhaskara Rao, Editor (2004), *Methods of Teaching Rural Sociology,* New Delhi: Discovery Publishing House, ISBN 81-7141-811-2.

Venkateswarlu, K. and S.J. Basha, Authors and Digumarti Bhaskara Rao, Editor (2004), *Methods of Teaching Commerce,* New Delhi: Discovery Publishing House, ISBN 81-7141-808-2.

Venugopala Rao, K., Author and Digumarti Bhaskara Rao, Editor (2000), *Teacher Morale in Secondary Schools,* New Delhi: Discovery Publishing House, ISBN 81-7141-551-2.

Venugopala Rao, K., Author and Digumarti Bhaskara Rao, Editor (2007), *Techniques of Teaching History,* New Delhi: Sonali Publications, ISBN 81-8411-059-6.

Vidya, C., Author and Digumarti Bhaskara Rao, Editor (1996), *A Text Book of Nutrition,* New Delhi: Discovery Publishing House, ISBN 81-7141-309-9.

Vimala, T.D., B. Prasad Babu and Digumarti Bhaskara Rao, Editors (2007), *Stress, Coping and Management*, New Delhi: Sonali Publications, ISBN 81-8411-086-3.

Vijaya Bharathi, D., Author and Digumarti Bhaskara Rao, Editor (2000), *Educational Philosophies of Swami Vivekananda and John Dewey,* New Delhi: APH Publishing House, ISBN 81-7648-309-9.

Vijaya Bharathi, D., Author and Digumarti Bhaskara Rao, Editor (2005), *Educational Philosophy of John Dewey,* New Delhi: Discovery Publishing House, ISBN 81-8356-024-5.

Vijaya Bharathi, D., Author and Digumarti Bhaskara Rao, Editor (2005), *Educational Philosophy of Swami Vivekananda,* New Delhi: Discovery Publishing House, ISBN 81-8356-023-7.

Vijaya Lakshmi, D., Author and Digumarti Bhaskara Rao, Editor (2004), *Basic Education,* New Delhi: Discovery Publishing House, ISBN 81-7141-881-3.

Vijaya Lakshmi, V., Author and Digumarti Bhaskara Rao, Editor (2006), *Techniques of Teaching Music,* New Delhi: Sonali Publications, ISBN 81-8411-038-3.

Vijaya Kumar, S.J., Author and Digumarti Bhaskara Rao, Editor (2006), *Techniques of Teaching Mathematics,* New Delhi: Sonali Publications, ISBN 81-8411-039-1.

Visalakshi, V., Author and Digumarti Bhaskara Rao, Editor (2006), *Techniques of Teaching Biology,* New Delhi: Sonali Publications, ISBN 81-8411-045-6.

Visalakshi, V., Author and Digumarti Bhaskara Rao, Editor (2007), *Techniques of Teaching Zoology,* New Delhi: Sonali Publications, ISBN 81-8411-055-3.

Bhaskara Rao, Digumarti (1986), *Dhrushya Sravana Bodhanapakaranalu* (Audio Visual Teaching Aids). Guntur: Nagarjuna Publishers.

Bhaskara Rao, Digumarti (1993), *Jeevasashtra Bodhana* (Teaching of Biology), Guntur: Nagarjuna Publishers.

Bhaskara Rao, Digumarti (1995), *Vignanasasthra Bodhana* (Teaching of Science) Guntur: Nagarjuna Publishers.

Bhaskara Rao, Digumarti (1997), *Vidya Manovignana Sastram* (Educational Psychology), Guntur: Creative Press.

Bhaskara Rao, Digumarti (1998), *DSC Study Material,* Guntur: Nagarjuna Publishers.

Bhaskara Rao, Digumarti (1998), *Upadhyayudu Vidya,* (Teacher and Education) Guntur: Nagarjuna Publishers.

Bhaskara Rao, Digumarti (1998), *Vidya Drukpadalu* (Perspectives of Education), Guntur: Nagarjuna Publishers.

Bhaskara Rao, Digumarti (1999), *EdCET Teaching Aptitude,* Guntur: Nagarjuna Publishers.

Bhaskara Rao, Digumarti (2001), *Bharata Samajamulo Upadyayudu Vidhya* (Teacher and Education in Emerging Indian Society), Guntur: Sri Nagarjuna Publishers.

Bhaskara Rao, Digumarti (2001), *Bhoutika Sastra Bodhana Padhatulu* (Methods of Teaching Physical Science), Guntur: Sri Nagarjuna Publishers.

Bhaskara Rao, Digumarti (2001), *Jeeva Sastra Bodhana Padhatulu* (Methods of Teaching Biology),Guntur: Sri Nagarjuna Publishers.

Bhaskara Rao, Digumarti (2001), *Vidya Manovignana Sastram* (Educational Psychology), Guntur: Sri Nagarjuna Publishers.

Bhaskara Rao, Digumarti (2003), *Patasala Yajamanyam / Paripalana* (School Management and Administration), Guntur: Sri Nagarjuna Publishers.

Gopala Krishna, G., A. Rama Krishna, K. Subba Rao and Bhaskara Rao, Digumarti (2004), *Jeevasashtra Bodhana Padhatulu* (Methods of Teaching of Biological Science), Guntur: Sri Nagarjuna Publishers.

Krishna Murthy, V., K.S. Sudheer Reddy and Digumarti Bhaskara Rao (2004), *Vidya Manovignana Sastra Adharalu* (Foundations of Educational Psychology), Guntur: Sri Nagarjuna Publishers.

Lalini, V., V. Dayakara Reddy, M. Srihari and Digumarti Bhaskara Rao (2004), *Vidya Adharalu* (Foundations of Education), Guntur: Sri Nagarjuna Publishers.

Subba Rao, K.P., P. Ayodhya and Digumarti Bhaskara Rao (2004), *Patasala Yajamanyam – Vidhya Vyavusthalu* (School Management and Systems of Education), Guntur: Sri Nagarjuna Publishers.

Sudhakar, V., B. Ravindra Babu, D.S. Kumar and Digumarti Bhaskara Rao (2004), *Vidya Sanketika Sastram – Computer Vidhya* (Educational Technology and Computer Education), Guntur: Sri Nagarjuna Publishers.

Index

□□□